Framework
SCIENCE

Foundations

Sarah Jagger

OXFORD
UNIVERSITY PRESS

OXFORD
UNIVERSITY PRESS

Great Clarendon Street, Oxford OX2 6DP

Oxford University Press is a department of the University of Oxford.
It furthers the University's objective of excellence in research, scholarship,
and education by publishing worldwide in

Oxford   New York

Auckland   Cape Town   Dar es Salaam   Hong Kong   Karachi
Kuala Lumpur   Madrid   Melbourne   Mexico City   Nairobi
New Delhi   Shanghai   Taipei   Toronto

With offices in

Argentina   Austria   Brazil   Chile   Czech Republic   France   Greece
Guatemala   Hungary   Italy   Japan   South Korea   Poland   Portugal
Singapore   Switzerland   Thailand   Turkey   Ukraine   Vietnam

Oxford is a registered trade mark of Oxford University Press
in the UK and in certain other countries

British Library Cataloguing in Publication Data

Data available

ISBN 0 19 915008 7

ISBN 978 019 915008 3

1 3 5 7 9 10 8 6 4 2

Printed in Italy by Rotolito Lombarda.

Acknowledgements

The Publisher would like to thank the following for permission to
reproduce photographs:

P8 Aerial Shelly/Corbis UK Ltd; **p9** Don Fawcett/Science Photo Library;
p10 Photodisc/OUP; **p11** Photodisc/OUP; **p12** Jeremy Hutton Hibbert/
Science Photo Library; **p17** OUP; **p18t&m** Zooid/OUP; **p18b** Photodisc/OUP;
p19 Pictor International/Image State/Alamy; **p26** Yann Arthus-Bertrand/
Corbis UK Ltd; **p31** Holt Studios International; **p32** Nigel Cattlin/Holt
Studios International; **p37** Martyn F Chillmaid; **p43bl** Andrew Lambert
Photography/Science Photo Library; **p43tr** Archive Iconagrafico, S.A. &
Jonathan Blair/Corbis UK Ltd; **p44** Andrew F Lambert Photography/Science
Photo Library; **p45** Martyn F Chillmaid; **p46** Andrew Lambert Photography/
Science Photo Library; **p47l** Paul Schermaidster/Corbis UK Ltd; **p47r**
Biophoto Associates/Science Photo Library; **p52** Martyn F Chillmaid; **p53tr**
Nick Jorgensen/Rex Features; **p53bl** Larry Lee Photography/Corbis UK Ltd;
p57 Photodisc/OUP; **p66** Photodisc/OUP; **p75** Photodisc/OUP; **p79** Seiko
Europe Ltd; **p80tr** Oxford University Press; **p80br** Duomo/Corbis UK Ltd;
p81 Photodisc/OUP; **p85** Photodisc/OUP.

Technical illustrations are by Oxford Designers and Illustrators.

Cartoons are by John Hallet.

Front cover photos: Photodisc

Contents

Introduction

Everyone in school does science because it helps us to make decisions about our everyday lives. Science is also fun, and by studying science we get to find out how things work. This book is designed to help you learn about the key ideas in science that are taught in Year 9. We hope that you enjoy it.

In Year 9 you take your SATs, tests which measure your ability against students in the rest of the country. You will be tested on the science that you have covered in Years 7, 8 and 9. You can use this book along with the Year 7 and 8 books to help you revise.

How to use this book

This book is divided into 12 topics:

O At the start of each topic you will find an **opener page**. This page will remind you of what you already know about a topic, and will introduce the key ideas that you are about to meet.

O At the bottom of each main page there are some **questions** for you to test your understanding. Most of them can be answered using the information on the page, but some will require you to use your thinking skills and apply what you have just learnt. These questions are indicated by a thought bubble like this one:

O At the end of each topic there is a **'What have I learnt?'** page, with questions for you to test yourself.

O If you want to find out about something in particular use the **Contents** or the **Index**.

O The **Glossary** explains what certain words mean.

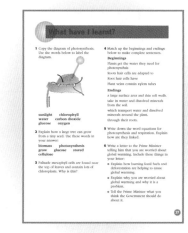

Inheritance and selection

What I should already know

- Cells are specialised to do different jobs.
- The nucleus of a plant or animal cell contains genetic information.
- Two parents are needed for sexual reproduction.
- Organisms of the same type are called a species.
- Organisms of the same species look different because of variation.

What I am going to meet in this unit

- Characteristics can be affected by inherited variation, environmental variation or both.
- Sexual reproduction produces offspring with characteristics from both parents.
- Some organisms have characteristics that help them survive, or that are useful to humans.
- Farmers can make new breeds of animals and plants by selective breeding.
- Only one parent is needed for sexual reproduction.

Asexual reproduction

Sexual reproduction

Organisms of the same type are called a **species**. The people in the picture all belong to the human species, but they don't all look the same. They have different **characteristics** because of **variation**.

There are two types of variation. Some characteristics like height are examples of both:

o **Inherited variation** is caused by the **genes** (genetic information) that we get from our parents. Eye colour, hair colour and nose shape are examples of inherited variation.

o **Environmental variation** is caused by an organism's environment. Factors like diet and illness affect how we look. Pierced ears and hair style are examples of environmental variation.

Some of the people in this family photo are related by blood (genetically). They look similar, but not the same because of variation.

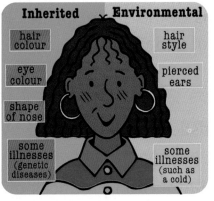

Examples of inherited and environmental variation.

1 What do we call organisms of the same type?

2 Unscramble these words:

traiviano   arescrithactsic   sneeg

3 Look at the eye colours of the other students in your class. Draw a tally chart to show the number of students with each eye colour. Is eye colour an example of environmental or inherited variation?

Copy and complete using these words:

**eye colour   variation   hair style
inherited   height   environmental**

There are two types of _____. Some characteristics like _____ are examples of both. _____ variation is caused by the genes that we get from our parents. _____ _____ is an example. _____ variation is caused by our environment. _____ _____ is an example.

Two parents are needed for **sexual reproduction**. For sexual reproduction to happen, a male and a female **gamete** (sex cell) must fuse (join) together. This is called **fertilisation**. Each gamete contains half of the genetic information needed to make a new life.

We look like our parents because we inherit our **genes** (these carry genetic information) from them, but there are also differences because of **inherited variation**. When a new baby is produced, half of the genetic information comes from the father (in the sperm) and half from the mother (in the egg). The child has characteristics from both parents because it has inherited a mixture of genes from them.

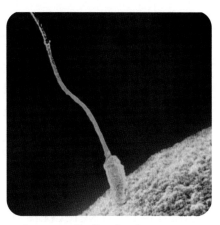

In humans, fertilisation happens when a sperm and an egg fuse.

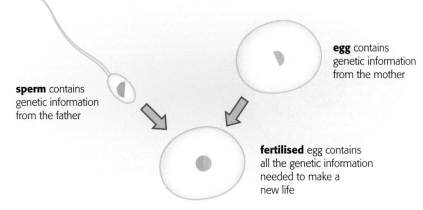

sperm contains genetic information from the father

egg contains genetic information from the mother

fertilised egg contains all the genetic information needed to make a new life

We inherit half our genes from each parent.

1 In humans, what are the male and female gametes called?

2 These words have had their vowels removed. What should they say?

frtlstn   gmt   gns

3 Draw and label a diagram like the one above to explain how we inherit characteristics from both our parents.

Copy and complete using these words:

characteristics   sexual reproduction genes   genetic information   gamete

Two parents are needed for _____ _____. Each _____ contains half of the _____ needed to make a new life. When a new baby is produced, half of the _____ _____ comes from each parent. The child will have _____ from both parents.

Natural selection

Organisms of the same species have different characteristics because of **variation**. This means that some individuals are more likely to survive than others. For example, an owl may have genes that make it silent when it flies. It will be able to hunt quietly and catch plenty of food. It is more likely to survive and produce offspring (that may also fly silently) than owls that make lots of noise when they fly. The characteristic of silent flight has been **selected**. This is called **natural selection**.

Owls have genes for silent flight.

Artificial selection

Some animals and plants have characteristics that are useful to people. People may then choose to breed from them, to produce offspring with the same characteristics. This is called **artificial selection** or **selective breeding**. Over many generations a new breed of animal or plant may be produced.

People use selective breeding to produce sheep that make lots of wool.

1 How does natural selection happen?
2 These words have had their vowels removed. What should they say?

   vrtn   slctn   ntrl   rtfcl

3 You have decided that you want to be a chicken farmer. You have bought some chickens that all have different characteristics. What characteristics might you want to select and breed for in your chickens?

Copy and complete using these words:
artificial   survive   useful
offspring   breeding   natural

In ____ selection, characteristics are selected because they make individuals more likely to ____ and produce ____. In ____ selection, people breed from animals or plants that have ____ characteristics. This is also called selective ____.

In **selective breeding** (artificial selection) people breed from animals or plants that have **desirable** (useful) **characteristics**. For example, a farmer wants to breed sheep that produce lots of wool and give a lot of meat (have lots of muscle). He decides to mate Bob and Polly because their offspring are likely to have these characteristics. Over many generations, a new breed of sheep may be produced.

Polly
produces wool but isn't very muscly.

Molly
isn't very wooly or muscly.

Bob
isn't very wooly but does have lots of muscle.

Selective breeding has made Dalmatians spotty, but they are often deaf.

There are some disadvantages of selective breeding:

o The characteristics that are selected are useful to humans, but may not be best for the plant or animal
o Some genetic problems can get selected by mistake
o Selective breeding reduces the amount of **variation**.

---

1 What is selective breeding?.

2 Unscramble these words:

   rasdibele   eltevsiec   oativarin

3 Why did the farmer decide to mate Bob with Polly and not with Molly? Explain your answer using the words: **desirable   characteristics   offspring**

Copy and complete using these words:

**variation   selective   desirable
characteristics   breeds**

In _____ breeding, people breed from animals or plants that have _____ _____. Over many generations, new _____ are produced but the amount of _____ in the population is reduced.

Two parents are needed for **sexual reproduction**. The offspring inherit a mixture of genes from the mother and the father. **Asexual reproduction** is different. Only one parent is needed, and the offspring has the same genes as the parent. We say that they are **clones**.

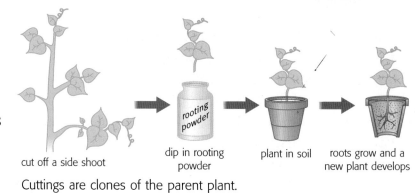

cut off a side shoot   dip in rooting powder   plant in soil   roots grow and a new plant develops

Cuttings are clones of the parent plant.

Plant growers often produce clones by taking cuttings from plants with **desirable characteristics**. They are able to produce identical plants quickly, but there are also disadvantages:

Advantages	Disadvantages
you can make identical copies of a really good plant	there's no variation in the population
it's quicker than waiting for seeds to grow	a disease could kill the whole population, not just a few plants

It is now possible to clone animals too, but it is very difficult. Many people think that it is wrong to clone animals and that it should not be done, especially on humans.

Dolly the sheep was the first animal to be cloned.

1 How many parents are needed in asexual reproduction?

2 What are the advantages and disadvantages of cloning plants?

3 Clones are genetically identical, but they may not always look the same. What environmental factors might affect the way that clones look?

Copy and complete using these words:

one   sexual   clone
inherit   asexual

In _____ reproduction the offspring _____ a mixture of genes from both parents. Only _____ parent is needed for _____ reproduction. The offspring has the same genes as the parent and is called a _____.

What have I learnt?

1 Write three lists. The first list should be of characteristics that are affected by inherited variation. The second list should be of characteristics that are affected by environmental variation. The third list should be of characteristics that are affected by both types of variation.

2 Our blood group is a characteristic that we inherit from our parents. The table below shows the percentage of the population in Britain with each blood group. Draw a bar chart to show this information.

Blood group	Percentage
A	40
B	11
O	45
AB	4

3 What is the difference between natural selection and artificial selection?

4 Match up the beginnings and endings below to make complete sentences.

Beginnings

In selective breeding, people breed from

A cattle farmer might pick two cows

Lettuce growers may try to produce

Selective breeding

Endings

lettuces with a nice flavour, that grow in poor soil.

plants or animals that have desirable characteristics.

reduces the amount of variation in a population.

that produce good quality meat and breed them together.

5 Explain how sexual reproduction and asexual reproduction differ. Include these words in your answer:

**offspring   parent   mixture
genes   clones**

6 A farmer is trying to produce a new breed of strawberries and has asked your advice. He has four different strawberry plants, each of which has different characteristics:

big red but not very juicy strawberries

small pink strawberries

big pink strawberries

small red juicy strawberries

The farmer wants to breed strawberry plants that produce large red juicy strawberries. Write a letter to the farmer telling him which two strawberry plants he should breed from. Explain why breeding from these two plants will produce offspring with the characteristics that he wants.

13

Fit and healthy

What I should already know

- Plants and animals are made up of cells, tissues, organs and organ systems.

- A developing fetus gets everything it needs from its mother's body through the placenta.

- We need to eat all of the seven nutrients to stay healthy.

- Animals and plants release energy from their food by respiration.

What I am going to meet in this unit

- What we mean by fitness.

- How exercise affects the body.

- What happens if we don't eat a balanced diet.

- How smoking and drinking affects our bodies.

- Different types of drugs and their effects on the body.

Fitness means different things to different people!

Fitness means different things to different people. The easiest way to think of fitness is that it is a measure of how well your body works. If you are fit, the **organs** and **organ systems** in your body work well together. You get your breath back quickly when you exercise, and you don't get too tired. The fitter you are, the faster your pulse rate returns to normal after exercise.

Some people need strong, muscular bodies. Other people need to be flexible.

There are many aspects to being fit. Different sports need different combinations of these.

To stay fit you need to look after your body. You can do this by eating a balanced diet, getting regular exercise and avoiding things that harm your body like drugs and smoking.

1 What do we mean by fitness?

2 These words have had their vowels removed. What should they say?

xrcs   rgns   blncd dt

3 Think about your favourite sport and then look at the different aspects of fitness in the picture above. Which of them do you need to be able to play your favourite sport?

Copy and complete using these words:

**pulse   drugs   fitness
balanced diet   exercise**

_____ is a measure of how well your body works. If you are fit, your _____ quickly returns to normal after you _____. To stay fit you need to look after your body by eating a _____ _____, exercising and avoiding _____ and smoking.

To be able to exercise we need **energy**. This energy is released from our food by a process called **respiration**:

glucose + **oxygen** → **carbon dioxide** + **water**   Energy is released

The oxygen that we need for respiration is provided by the **respiratory system**, which includes the lungs. When you breathe in, your ribs move up and out and your **diaphragm** (a sheet of muscle) flattens. This causes air to rush into your lungs. Oxygen from the air passes into the blood and is transported around the body by the **circulatory system**. The fitter you are, the greater the volume of air that your lungs can hold.

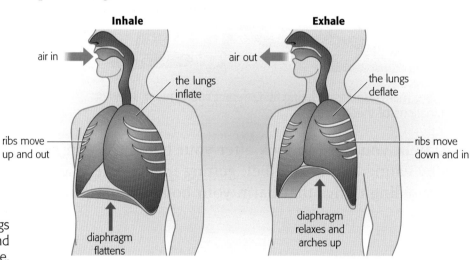

When you breathe out, your diaphragm relaxes and your ribs move down and in. This forces air out of your lungs. Carbon dioxide that has passed into the lungs from the blood leaves the body.

Ventilation (breathing) brings oxygen into our bodies and removes carbon dioxide.

1 Write down the word equation for respiration.

2 Name the two organ systems that provide our cells with the oxygen that they need for respiration.

3 Draw two flow diagrams. The first one should explain how oxygen from the air gets to our cells for respiration. The second should explain how the carbon dioxide produced by respiration leaves our bodies.

Copy and complete using these words:

**respiratory   energy   removes
diaphragm   respiration   ventilation**

_____ is released from our food by _____. The oxygen that is needed for respiration is provided by the _____ system. It also _____ carbon dioxide from the body. _____ (breathing) is caused by the movement of the ribs and _____.

You are what you eat

To be healthy, we need to eat a **balanced diet**. This means that we need to eat the right amounts of the seven main **nutrients**: protein, carbohydrate, fat, fibre, vitamins, minerals and water.

If your diet is not balanced, your body can't work properly. For example, if a certain vitamin or mineral is missing from your diet you might develop a **deficiency disease**. The table shows some examples.

In the same way, you can become ill if your diet contains too much fat and sugar. More and more people are becoming **obese** (very overweight) because they do not eat a balanced diet or get enough exercise. Obesity is a serious problem because it can cause conditions like diabetes and heart disease.

Too many sugary or fatty foods can stop your body from working properly.

These conditions can all be caused by obesity.

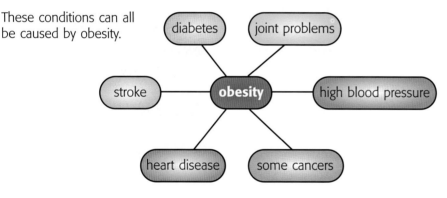

Vitamin/ mineral	Deficiency disease
Vitamin C	Scurvy
Vitamin D	Rickets
Iron	Anaemia
Calcium	Osteoporosis

1 What are the seven nutrients that we need to eat to have a balanced diet?

2 Unscramble these deficiency diseases:

 eiamana   vruysc   kercist

3 More and more people are becoming obese. Explain what this means and why it is a serious problem. You should include a list of some of the conditions that can be caused by obesity.

Copy and complete using these words:

diabetes   deficiency   balanced diet   obese   nutrients

To be healthy we need to eat a _____ _____. We need to eat the right amounts of the seven main _____. If a nutrient is missing from our diet we might develop a _____ disease. If we become _____ we can develop conditions like _____ and heart disease.

Drugs affect the way the body or the brain works. Many drugs are **addictive**. People may feel that they need to keep taking them, and may have **withdrawal symptoms** if they stop. Most drugs also have **side effects** (effects on the body other than the ones we want). There are three types of drug:

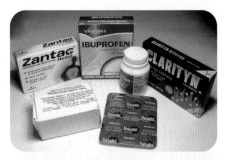

o **Medicinal drugs** are used to treat people who are ill. They can be bought from a pharmacist or prescribed by a doctor. Examples include paracetamol and antibiotics.

o **Recreational drugs** are drugs that people take because they like the effect the drug has. Examples include alcohol, caffeine in coffee and nicotine in cigarettes.

o **Illegal drugs** are drugs that people take because they like the effect, but they can be very harmful. That's why they're illegal. Examples include cannabis, heroin and ecstasy.

1 What are drugs?
2 Name the three different types of drugs and give two examples of each type.
3 Illegal drugs can be very harmful, but people still take them. Try to think of as many reasons as you can why people take illegal drugs, even though they know that they may do them harm.

Copy and complete using these words:

effects   illegal   addictive
recreational   brain   medicinal

Drugs affect the way the body or the _____ works. Many are _____ and most of them have side _____. _____ drugs are used to treat illnesses. People take _____ and illegal drugs because they like the effect, but _____ drugs can be very harmful.

Alcohol and smoking

Alcohol

Alcohol is a **recreational drug**. A little alcohol can make you feel relaxed and happy, but too much can make you dizzy, or even unconscious. Alcohol also slows down your reactions, which is why you should never drive after drinking. People can become **addicted** to alcohol. Drinking too much alcohol too often can also damage your liver and cause brain damage.

Smoking

Smoking is addictive and damages the **respiratory system**. Cigarette smoke contains lots of poisonous substances, including carbon monoxide, nicotine and tar. The heat from the hot smoke also damages your airways and lungs.

Women who drink or smoke while pregnant risk harming their unborn babies. Poisonous substances pass from the mother to the fetus through the placenta.

Substance	What does it do?
Carbon monoxide	Makes cells in your blood less able to carry oxygen around the body.
Nicotine	An addictive drug. Causes high blood pressure, which can lead to heart disease.
Tar	Coats the inside of the lungs. Also contains substances that cause cancer.

Some substances in cigarette smoke.

1 How can drinking too much alcohol damage your body?

2 Unscramble these words:

ronbac domenixo   tencoini   rat

3 Name three poisons that are found in cigarette smoke. Write down the effect that each one has on the body. How can an unborn baby be harmed by these poisons if its mother smokes?

Copy and complete using these words:

liver   addictive   nicotine respiratory   reactions   alcohol

_____ and smoking are _____. Alcohol slows down your _____. Drinking too much too often can damage your _____ and cause brain damage. Smoking harms the _____ system. Cigarette smoke contains poisons like carbon monoxide, tar and _____.

What have I learnt?

1 Krupa is a good swimmer. Her ambition is to compete in the Olympics. What should Krupa do to make sure that she is fit? Which aspects of fitness will she need to develop to win a gold medal?

2 Copy and complete the table below by putting the phrases in the list into the correct columns.

Breathing in	Breathing out

lungs inflate

ribs move down and in

lungs deflate

air is drawn into the lungs

air is forced out of the lungs

ribs move up and out

diaphragm relaxes

diaphragm flattens

3 Match up the beginnings and endings below to make complete sentences.

Beginnings

To be healthy you need to eat

Your body can't work properly

If there is not enough iron your diet

If you are obese

Endings

you are at risk of developing heart disease and diabetes.

if you don't eat a balanced diet.

the right amounts of the seven nutrients.

you could develop a deficiency disease called anaemia.

4 Sort the drugs below into three groups: medicinal, recreational and illegal.

**alcohol   heroin   aspirin   speed
nicotine   cough syrup   caffeine
cocaine   penicillin   cannabis
paracetamol   ecstacy**

5 You shouldn't drive after drinking alcohol. Why does drinking alcohol make it unsafe for you to drive?

6 Write a newspaper article about illegal drugs. Name some examples of illegal drugs in your article and try to answer these questions:

o Why is it against the law to take illegal drugs?

o How do they affect the body?

o Why do people use them?

o What are the dangers of taking them?

Plants and photosynthesis

What I should already know

- Living things can be sorted into groups.
- Living things are adapted to survive in their habitats.
- Plants are producers. They use energy from the Sun to make their food.
- How to write a word equation to describe a chemical reaction.

What I am going to meet in this unit

- Green plants make their own food by photosynthesis.
- In photosynthesis, the reactants are carbon dioxide and water. The products are glucose and oxygen.
- Leaves are adapted for photosynthesis.
- Roots are adapted for taking in water.
- Respiration and photosynthesis are linked.
- Why plants are important to the environment.

And for my next trick ...

YEARS LATER...

No. Just a gain in biomass.

Wow! Was that magic?

Plants are **producers**. They make their own food by a process called **photosynthesis**. In photosynthesis, plants turn water (from the soil) and carbon dioxide (from the air) into glucose and oxygen. To do this, they need a green chemical called chlorophyll. The also need energy from the Sun. The Sun's energy is transformed into chemical energy in glucose.

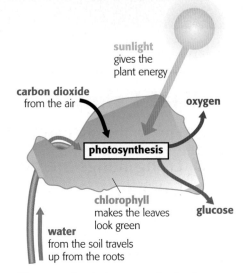

Photosynthesis happens in the leaves of green plants.

We can write a word equation to describe photosynthesis. The reactants are water and carbon dioxide. The products are glucose and oxygen. We also need to include sunlight and chlorophyll because photosynthesis can't happen without them.

sunlight

carbon dioxide + water → glucose + oxygen

chlorophyll

We can measure the rate of photosynthesis by shining a light on pondweed in water. Bubbles of oxygen form in the water as the weed photosynthesises.

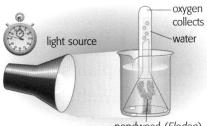

Oxygen collects in the tube as the pondweed photosynthesises. You can then measure the volume of gas produced in a certain time.

1 What is the name of the process that plants use to make their food?

2 Which of the following substances are produced by photosynthesis?

**oxygen   carbon dioxide
chlorophyll   glucose   water**

3 If you leave a plant in a dark place for a few days its leaves will turn pale. Explain why this happens.

Copy and complete using these words:

**chlorophyll   producers   water
energy   photosynthesis   glucose**

Plants are _____. They make their own food by _____. Carbon dioxide and _____ are turned into _____ and oxygen. To do this, they need _____ from the Sun and a green chemical called _____.

How does a large tree grow from a tiny seed? The answer is by gaining **biomass** (the mass of all the material in a plant or animal).

Like all living things, plants need to release energy by **respiration**. They use some of the oxygen and glucose that they make by **photosynthesis** for this, but there is still plenty of glucose left over. Rather than waste it, plants turn the left over glucose into other substances. Some of the glucose is stored for later as **starch**, and some is turned into substances that they can use to grow. The more they do this, the greater their biomass and the bigger they get!

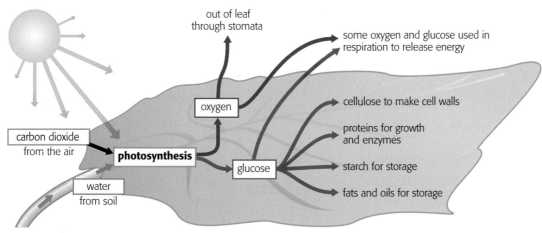

out of leaf through stomata

some oxygen and glucose used in respiration to release energy

oxygen

cellulose to make cell walls

carbon dioxide from the air

photosynthesis

proteins for growth and enzymes

glucose

starch for storage

water from soil

fats and oils for storage

---

1 What do we mean by biomass?

2 Unscramble these words:

sriteiparon   somibas   thishtonesopsy

3 Plants make glucose and oxygen by photosynthesis. Write a list of the different things that plants use the glucose for. What is some of the oxygen used for?

Copy and complete using these words:

glucose   starch   biomass photosynthesis   respiration

Plants make glucose and oxygen by _____. Some of it is used for _____. The _____ that is left over is turned into other substances like _____ for storage. As a result, the plant gains _____ and grows.

Like animals, plants have **organs**. Leaves are the organs where photosynthesis happens, and they are specially adapted to do their job. Leaves are wide and flat so that they catch lots of sunlight. The cells near the top have lots of **chloroplasts** (these are where **chlorophyll** is stored) to trap sunlight. There are also holes at the bottom of leaves called **stomata** (stoma if you're only talking about one) where carbon dioxide can enter, and oxygen can leave.

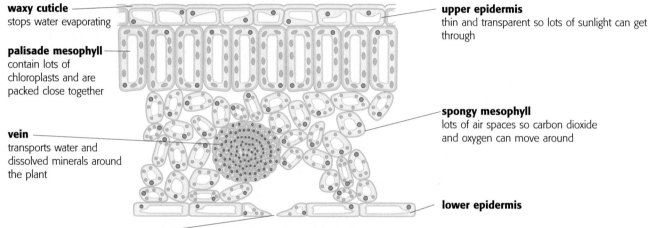

waxy cuticle
stops water evaporating

palisade mesophyll
contain lots of chloroplasts and are packed close together

vein
transports water and dissolved minerals around the plant

upper epidermis
thin and transparent so lots of sunlight can get through

spongy mesophyll
lots of air spaces so carbon dioxide and oxygen can move around

lower epidermis

stoma lets gases get into and out of the leaf

We can show that chlorophyll is needed for photosynthesis by testing a variegated (green and white) leaf with iodine. Only the green parts of the leaf turn blue-black, showing that starch is present.

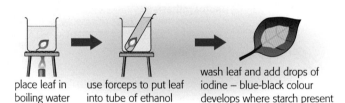

place leaf in boiling water

use forceps to put leaf into tube of ethanol

wash leaf and add drops of iodine – blue-black colour develops where starch present

1 How are leaves adapted to do their job?

2 You could test a leaf for starch. Write notes explaining how.

3 Draw a diagram of the structure of a leaf. Add labels to explain how each part of the leaf helps it to do its job.

Copy and complete using these words:

**sunlight   stomata   organs
chloroplasts**

Leaves are the plant _____ where photosynthesis happens. They have holes called _____ and cells with lots of _____. They are also shaped to trap lots of _____.

Both **carbon dioxide** and **water** are needed for photosynthesis. Unlike carbon dioxide, plants can't get the water they need through their leaves. Instead, they get it from the soil through their **roots**.

Like leaves, roots are plant organs that are adapted to do their job. They do two important things:

o absorb water and dissolved minerals

o anchor the plant in the soil.

As a plant grows, its roots branch out into the soil. They are covered in **root hair cells** that are adapted to take in lots of water. They have very thin cell walls, and give the roots a large surface area. Once the water and dissolved minerals have been taken in, they are transported around the plant by **xylem** and **phloem** tubes in the veins.

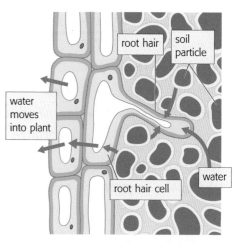

Root hair cells have a large surface area so that they can absorb lots of water.

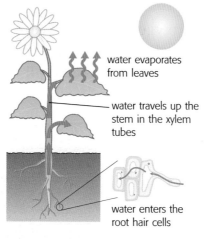

water evaporates from leaves

water travels up the stem in the xylem tubes

water enters the root hair cells

Plants have a circulatory system too! Water and dissolved minerals are carried around the plant by xylem and phloem tubes in the veins.

---

1 How do plants get the water they need for photosynthesis?

2 These words have had their vowels removed. What should they say?

mnrls   xylm   phlm   wtr

3 Like all other cells, root hair cells release energy by respiration. They can die if the soil that they are growing in becomes waterlogged. Explain why.

Copy and complete using these words:

**xylem   adapted   water   veins
minerals   surface**

Plants get the _____ and dissolved _____ they need through their roots. _____ and phloem tubes in the _____ then transport them around the plant. Root hair cells are _____ to absorb lots of water. They have a large _____ area and very thin cell walls.

Look at the word equations for photosynthesis and respiration. They are linked:

Photosynthesis

$$\text{carbon dioxide + water} \xrightarrow[\text{chlorophyll}]{\text{sunlight}} \text{glucose + oxygen}$$

Respiration

glucose + oxygen → carbon dioxide + water   energy is released

This mouse has everything it needs: water, food and oxygen.

All living things need glucose and oxygen for **respiration**. Both are produced by **photosynthesis**. This means that plants are very important. Without them, we would all die because we would not have any food or oxygen.

This mouse is in trouble. Without plants it has no oxygen to breathe.

Things are balanced if the amount of carbon dioxide produced by respiration is the same as that used for photosynthesis, but we know that the level of carbon dioxide in the atmosphere is rising. Burning **fossil fuels** (produces carbon dioxide) and **deforestation** (cutting down lots of trees) is helping to cause **global warming**. If this continues we will all be in big trouble.

Deforestation contributes to global warming.

1 Without plants we would all die. Explain why.

2 Unscramble these words:

   retadosfniot   laglob rangwim gyonex

3 What does deforestation mean? How is it helping to cause global warming?

Copy and complete using these words:

oxygen   glucose   deforestation photosynthesis   respiration

Respiration and _____ and linked. Plants produce the _____ and _____ that all living things need for _____. _____ is helping to cause global warming.

What have I learnt?

1 Copy the diagram of photosynthesis. Use the words below to label the diagram.

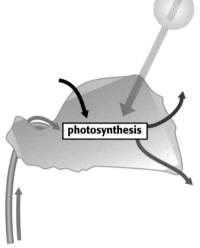

photosynthesis

sunlight   chlorophyll
water   carbon dioxide
glucose   oxygen

2 Explain how a large tree can grow from a tiny seed. Use these words in your answer:

biomass   photosynthesis
grow   glucose   stored
cellulose

3 Palisade mesophyll cells are found near the top of leaves and contain lots of chloroplasts. Why is this?

4 Match up the beginnings and endings below to make complete sentences.

Beginnings

Plants get the water they need for photosynthsis

Roots hair cells are adapted to

Root hair cells have

Plant veins contain xylem tubes

Endings

a large surface area and thin cell walls.

take in water and dissolved minerals from the soil.

which transport water and dissolved minerals around the plant.

through their roots.

5 Write down the word equations for photosynthesis and respiration. Explain how are they linked.

6 Write a letter to the Prime Minister telling him that you are worried about global warming. Include these things in your letter:

o Explain how burning fossil fuels and deforestation are helping to cause global warming.

o Explain why you are worried about global warming and why it is a problem.

o Tell the Prime Minister what you think the Government should do about it.

Plants for food

What I should already know

- The seven life processes.
- How to draw food chains and food webs.
- How to collect data about a habitat.
- The biomass of a plant is made by photosynthesis.
- The importance of photosynthesis.
- Plants need water, sunlight, carbon dioxide and minerals from the soil.

What I am going to meet in this unit

- Photosynthesis produces all our food, and the oxygen we need for respiration.
- We put fertilisers on soil so that plants have all the minerals that they need to grow.
- Weeds compete with crops.
- Pests compete with humans for food.
- Weedkillers and pesticides.

I don't think we need to put any more fertiliser on the carrots!

Food chains show who eats who. The arrows show how energy is **transferred** from one organism to another. All of the energy in food chains originally came from the Sun.

In the picture, Del is about to eat a beef and tomato sandwich. Here are some food chains for his sandwich:

grass ➡ cow (beef) ➡ human

wheat (bread) ➡ human

tomato ➡ human

All food chains start with a **producer**. Green plants are producers because they use the Sun's energy to make their own food by **photosynthesis**. Animals are **consumers** because they consume (eat) other organisms. There are three types of consumer:

o **herbivores** only eat plants

o **carnivores** only eat animals

o **omnivores** eat both plants and animals.

All of the energy in Del's beef and tomato sandwich originally came from the Sun.

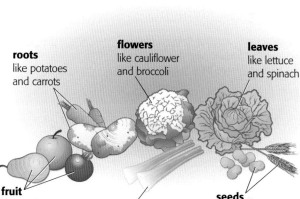

roots
like potatoes and carrots

flowers
like cauliflower and broccoli

leaves
like lettuce and spinach

fruit
like apples, pears and tomatoes

stems
like celery

seeds
like wheat (bread) and beans

We can eat many different parts of plants: stems, seeds, roots, leaves and even flowers!

1 Where does the energy in a food chain come from?

2 What are the three types of consumer?

3 Plants make their own food. Explain how they do this, and how they are able to grow. Use these words in your answer:

energy   Sun   biomass photosynthesis

Copy and complete using these words:

**omnivores   consumers   Sun
carnivores   energy   producer**

The _____ in all food chains originally came from the _____. Food chains always start with a _____. Animals that consume producers are called _____. The three type of consumer are: _____, herbivores and _____.

To be able to grow healthily, plants need small amounts of certain **minerals**. They take these up from the soil through their **roots**. Over time, the minerals in the soil run out. We can replace them by adding **fertiliser** to the soil. Farmers use fertilisers because they help them get the **maximum yield** from their crops (the largest amount of crop for every square metre of land).

Adding fertilisers to soil helps plants to grow healthily.

The three most important elements that plants need are **nitrogen** (N), **phosphorus** (P) and **potassium** (K). Without them, plants can't grow healthily. Farmers often add NPK fertiliser to their soil because it contains all three of these elements.

Element	Needed for		If absent
Nitrogen	Healthy leaves		Yellow/pale green leaves
Phosphorus	Healthy roots and flowers		Purple leaves and small roots
Potassium	Good general health, flowers and fruit		Yellow leaves with dead spots

1 How do plants obtain the minerals that they need?

2 These words have had their vowels removed. What should they say?

ntrgn   phsphrs   ptssm

3 Explain why farmers use fertilisers on their crops. You should be able to give at least two reasons.

Copy and complete using these words:

**yield   nitrogen   minerals
potassium   fertiliser**

To grow healthily, plants need to take up _____ from the soil like _____, phosphorus and _____. We can put minerals back into soil by adding _____. Fertilisers help farmers get the maximum _____ from their crops.

Organisms that live in a habitat **compete** with each other for resources like food. **Weeds** are plants that grow where we don't want them to. Any plant can be a weed, depending on where it grows. Weeds are bad news for farmers because they compete with crops for space, water, light and minerals in the soil. They can reduce the **crop yield**.

To get rid of weeds, farmers use chemicals called **herbicides** (weedkillers). There are two types of herbicide:

o **Selective herbicides** only kill certain weeds. If you spray them on a cornfield, the weeds will die but the corn will survive.

o **Non-selective herbicides** kill a wide range of plants. A farmer might use them to clear a whole area of plants before planting a new crop.

Farmers can spray herbicides over large areas.

1 What is a weed? Can flowering plants like daffodils be called weeds?

2 Unscramble these words:

mopecet   teecvelsi   crisibedeh

3 There are two types of herbicide. What are they and how do they work? For each type, describe what a farmer might use it for.

Copy and complete using these words:

herbicides   weeds   non-selective
light   compete   space

Organisms _____ with each other for resources. _____ compete with crops for _____, water, _____ and minerals in the soil. We can use chemicals called _____ to kill weeds. They can be either selective or _____.

Pests are animals that compete with humans for food. Mice, slugs and aphids (greenfly) are all pests that can damage crops and reduce the **crop yield**. We can use chemicals called **pesticides** to kill pests.

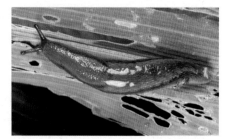
Slugs are pests that eat the leaves of plants.

Pesticides contain poisonous substances called **toxins**. They kill pests, but may also harm other organisms. Using pesticides can also affect the **populations** of other organisms in a food web. In the food chain below we can see that aphids attack roses. If we kill the aphids with a pesticide, the ladybirds might not be able to find enough food. If this happens, the number of ladybirds will drop. The number of robins may then also be affected.

rose ➡ aphid ➡ ladybird ➡ robin

Organic farmers try to control pests and weeds without using herbicides and pesticides. They might dig up weeds by hand, or introduce a predator of a pest to try to keep its numbers down.

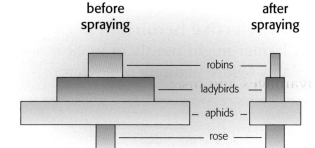
before spraying   after spraying

robins
ladybirds
aphids
rose

Pest control can affect the populations of other organisms in a food web.

---

1 What are pests? Give two examples of pests.

2 What is an organic farmer? How might an organic farmer try to control pests and weeds.

3 Look at the food chain below. What might happen to the population of owls if a farmer used a pesticide to kill the mice?

wheat ➡ mouse ➡ owl

Copy and complete using these words:

**yields   populations   crop   web
pests   pesticides**

_____ are animals that compete with humans for food. They can reduce _____ _____, so farmers use chemicals called _____ to kill them. Pesticides can affect the _____ of other organisms in the food _____. They can also harm other organisms.

To grow the **maximum yield** of a healthy crop, farmers need to provide the right **environmental conditions**. Their plants need plenty of light, water and carbon dioxide for **photosynthesis**, and minerals from the soil. Farmers also need to stop weeds from competing with their crops, and control pests and diseases.

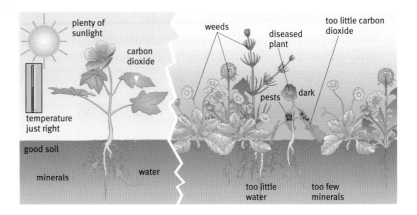

Some farmers grow their crops in **greenhouses** so that they can control the conditions that their crops grow in. Crops that only grow in the summer can be grown all year round. The farmer can adjust the temperature and the levels of light and carbon dioxide. Water and fertiliser can also be added when needed. There are advantages and disadvantages to growing crops this way:

Advantages of greenhouses	Disadvantages of greenhouses
Farmers can control the conditions that their crops grow in so that they get maximum yields.	It's expensive to grow crops this way so it's only used to grow crops that will fetch a high price, like strawberries.
It's easier to control pests and weeds.	If a pest gets established it can spread quickly and kill the whole crop.

1 What do plants need if a farmer is to get the maximum yield of a healthy crop?

2 Unscramble these words:

diely   ditniocsno   hunesergoes

3 Some farmers grow their crops in greenhouses. What are advantages of growing crops in large greenhouses? What are the disadvantages?

Copy and complete using these words:

environmental   yield   light   greenhouses   control   conditions

To grow the maximum _____ of a healthy crop, farmers need to grow their plants in the right _____ _____. Some farmers grow their crops in _____ so that they can _____ these conditions. They can adjust levels of _____ and carbon dioxide.

What have I learnt?

1 Copy and complete the table to show which parts of a plant we are eating when we eat each food. Choose from the list below.

Food	Part of plant
almond	
orange	
leek	
lettuce	
potato	

leaves　　**seed**　　**stem**　　**root**　　**fruit**

2 Jenny has been given three plants that aren't growing healthily. Which element does each plant need?

a The first plant has got purple leaves.

b The second plant has got yellow leaves with dead spots on them.

c The third plant also has some yellow leaves, but the leaves at the top of the plant are pale green.

3 When weeds grow in crop fields the crop yields are lower. Why is this?

4 Match up the beginnings and endings below to make complete sentences.

Beginnings

Pests are animals that

Farmers use chemicals called pesticides

Pesticides contain toxins that

Pesticides can affect

Endings

can harm other organisms in the food web.

compete with humans for food.

the populations of other organisms in the food web.

to kill pests.

5 Imagine that you are going to try and grow a crop of strawberries in a greenhouse. What conditions will you need to control? What are the advantages of growing strawberries in this way? What are the disadvantages?

6 You have been asked to write some questions for a new quiz programme on TV called 'Plants for Food'. Write five questions about what you have learnt in this unit. Don't forget to include the answers.

Reactions of metals and metal compounds

What I should already know

- About atoms, elements and compounds.
- We can write formulae to describe compounds.
- When a chemical reaction happens, atoms join together in new ways.
- Neutralisation reactions cancel out acids.
- How to test for hydrogen, oxygen and carbon dioxide.

What I am going to meet in this unit

- Signs that can tell us that a chemical reaction is taking place.
- A gas is produced when an acid reacts with a metal or a metal carbonate.
- What happens when acids react with metal oxides and alkalis.
- How to identify patterns in chemical reactions, and write word equations to describe them.
- Examples of salts that are formed during chemical reactions.

THE AMAZING
METAL FAMILY

KILO GRAMS

dense   strong

shiny

flexible

The **elements** in the periodic table can be split into two groups. **Metals** are grouped on the left-hand side of the periodic table. **Non-metals** are grouped on the right-hand side. The general properties of metals and non-metals are shown in the table below.

				H													He
Li	Be											B	C	N	O	F	Ne
Na	Mg											Al	Si	P	S	Cl	Ar
K	Ca	Sc	Ti	V	Cr	Mn	Fe	Co	Ni	Cu	Zn	Ga	Ge	As	Se	Br	Kr
Rb	Sr	Y	Zr	Nb	Mo	Tc	Ru	Rh	Pd	Ag	Cd	In	Sn	Sb	Te	I	Xe
Cs	Ba	La	Hf	Ta	W	Re	Os	Ir	Pt	Au	Hg	Tl	Pb	Bi	Po	At	Rn
Fr	Ra																

Metals are grouped in the blue area of the periodic table. Non-metals are grouped in the red area.

Metals	Non-metals
conduct heat and electricity well	do not conduct heat and electricity well
hard and strong	softer and weaker than metals
flexible (can be shaped)	brittle (may shatter if you hit them)
shiny (though some tarnish)	usually dull
high melting and boiling points	low melting and boiling points
dense (heavy for their size)	not very dense
solids at room temperature (except mercury which is a liquid)	gases at room temperature (except bromine which is a liquid)

Metal	What is it used for?
iron	building bridges (it's strong and cheap)
copper	electrical wiring (it's flexible and a good conductor of electricity)
titanium	racing bicycles (light and strong, but also very expensive)

The properties of metals make them useful for lots of different things.

1 Elements can be split into two groups. What are they?

2 Unscramble these words:

nemelets   letsam   roptiperes

3 Plastics are made in factories. They can be hard and strong, but they are not metals. They aren't non-metals either. Explain why.

Copy and complete using these words:

mercury   conductors   non-metals
solids   elements   left-hand

_____ can be split into two groups: metals and _____. Metals are found on the _____ side of the periodic table and are _____ at room temperature, except _____. They are hard, strong and good _____ of heat and electricity.

Metals and acids

Zinc is a **metal**. When zinc reacts with hydrochloric acid a salt (zinc chloride) and hydrogen are produced:

zinc + hydrochloric acid → zinc chloride + hydrogen

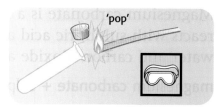

We can test for hydrogen with a lit splint. If hydrogen is present we'll hear a 'squeaky pop'.

You can tell that a reaction is taking place because:

o bubbles of gas are produced

o the piece of zinc gets smaller

o a new material is left behind if you evaporate the liquid left at the end of the reaction.

Other metals react with acids like zinc does, but not all metals. **Reactive** metals like magnesium bubble violently in acids, but less reactive metals like copper do not.

calcium   magnesium   zinc   iron   lead   copper

We can write this general word equation to describe what happens when a metal reacts with an acid:

metal + **acid** → salt + **hydrogen**

1 Which gas is produced when a metal reacts with an acid?

2 What three signs can tell us that a metal is reacting with an acid?

3 Write a word equation to describe what happens when calcium reacts with hydrochloric acid.

Copy and complete using these words:

metal   hydrogen   salt   smaller

When a _____ reacts with an acid a _____ and hydrogen gas are produced. The piece of metal gets _____ and bubbles of _____ are produced.

Most metals react with water, but **unreactive** metals like gold do not. Metals that react with water do not all react in the same way. Potassium is very **reactive** and bursts into flames when it reacts with water. Iron is less reactive, and **corrodes** slowly over time.

When a metal reacts with water, hydrogen gas and a **metal hydroxide** are produced. We can write a general word equation to describe what happens:

metal + water → metal hydroxide + hydrogen

For example, lithium hydroxide and hydrogen are produced when lithium reacts with water:

lithium + water → lithium hydroxide + hydrogen

We can show that hydrogen is produced by collecting the gas that is produced and doing the 'squeaky pop' test. The lithium hydroxide (metal hydroxide) makes the water alkaline.

Potassium bursts into flames when it reacts with water.

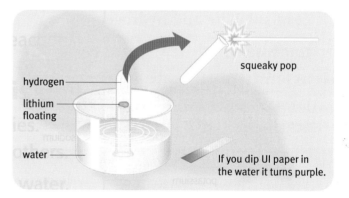

hydrogen

lithium floating

water

squeaky pop

If you dip UI paper in the water it turns purple.

1 Which gas is formed when a metal reacts with water?

2 Unscramble these words:

atmle xodehidry   cevritae

3 What are the products that are made when sodium (a metal) reacts with water? Write a word equation to describe the reaction.

Copy and complete using these words:

**pop   metal   hydrogen
hydroxide   water   squeaky**

Most metals react with _____, but they do not all react in the same way. When a metal reacts with water, hydrogen and a _____ _____are produced. The _____ _____ test tells us that one of the products is _____.

Magnesium burns in air because it reacts with oxygen in the air. A metal oxide called magnesium oxide is produced. A new material is formed, and heat and light energy are given out.

Most metals react with acids, water and oxygen. The general word equations below describe what happens in these reactions:

metal + **acid** → salt + **hydrogen**

metal + **water** → **metal hydroxide** + **hydrogen**

metal + **oxygen** → **metal oxide**

Metals react with acids, water or oxygen in different ways because some metals are more **reactive** (aggressive) than others. Reactive metals like potassium react quickly and violently. Less reactive metals like iron react more slowly. **Unreactive** metals like gold do not react at all.

The **reactivity series** lists metals in order of how reactive they are. The most reactive metals go at the top. Unreactive metals go at the bottom, below hydrogen. We can also use the reactivity series to make predictions. For example, we can predict that magnesium is more reactive than silver because it is higher up the reactivity series.

Reactivity series

potassium (K)
sodium (Na)
calcium (Ca)
magnesium (Mg)
aluminium (Al)
zinc (Zn)
iron (Fe)
lead (Pb)
(hydrogen (H))
copper (Cu)
silver (Ag)
gold (Au)

---

1 Write a general word equation to describe what happens when a metal reacts with oxygen.

2 These words have had their vowels removed. What should they say?

rctvty srs   **nrctv**

3 Metals that are less reactive than hydrogen are unreactive. Which metals are unreactive?

Copy and complete using these words:

series   reactive
unreactive   reactivity

_____ metals like potassium react quickly and violently. _____ metals like gold do not react with water, oxygen or acids. The _____ _____ lists metals in order of their reactivity.

We can use the **reactivity series** to predict what will happen in a chemical reaction. A more reactive metal will **displace** (knock out) a less reactive metal from a compound. For example, magnesium displaces copper from copper sulphate because magnesium is more reactive than copper:

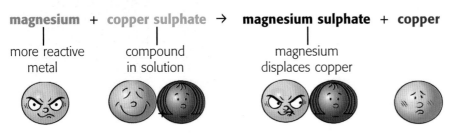

magnesium + copper sulphate → **magnesium sulphate** + **copper**

more reactive metal | compound in solution | magnesium displaces copper

Magnesium displaces copper from blue copper sulphate solution. Colourless magnesium sulphate solution forms, and the displaced copper falls to the bottom of the tube.

We can think of reactive metals as being better at holding onto things than less reactive metals. Magnesium holds onto the sulphate better than copper, so copper is displaced and left on its own.

magnesium   sulphate   copper

1 What happens in a displacement reaction?

2 These words have had their vowels removed. What should they say?

rctvty   dsplcmnt   mtl

3 Zinc is more reactive than iron. Write a word equation to describe what happens when zinc reacts with iron sulphate.

Copy and complete using these words:

**less   series   reaction   more
holding   reactivity**

We can use the _____ _____ to predict what will happen in a _____. A _____ reactive metal will displace a _____ reactive metal from its compound. Reactive metals are better at _____ onto things than less reactive metals.

The best metal for the job

We use different metals for different jobs because they have different **properties**. These properties depend on their reactivity. Reactive metals like sodium and potassium are too reactive for everyday use. Iron and aluminium are used to make cars because they are strong, can be shaped easily and aren't very reactive. Gold is unreactive, but it is also very expensive.

Imagine having a car that was made of potassium. It would burst into flames when it rained!

Unreactive metals like gold and silver are found in the ground as pure **metal elements**, not as compounds. Reactive metals are different. They are found in compounds called **ores**, because they have reacted with minerals in the ground. The more reactive a metal, the more difficult it is to extract it from its ore.

Gold can be found as an element in the ground.

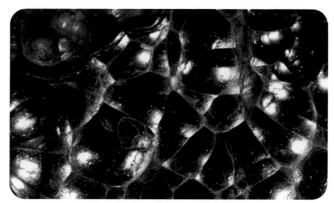

We have to extract iron from ores like this one.

1 Why aren't sodium and potassium used for everyday items?

2 Why are aluminium and iron used to make cars?

3 How do we find unreactive metals in the ground? How do we find reactive metals in the ground? Explain why it is difficult to extract potassium from its ore.

Copy and complete using these words:

**ores   properties   unreactive
reactive   metals   reactivity**

We use different _____ for different jobs because they have different _____. These properties depend on their _____. _____ metals are found in the ground as compounds called _____. _____ metals are found as pure metal elements.

1 a What do we mean when we say that a metal has corroded?

  b What do we mean when we say that a metal has tarnished?

2 Match up the beginnings and endings below to make complete sentences.

Beginnings

Most metals react with water to form

Metal hydroxides make

The 'squeaky pop' test tells us

When potassium and water react,

Endings

water alkaline.

a metal hydroxide and hydrogen.

potassium hydroxide and hydrogen are produced.

that hydrogen is produced.

3 Look at the reactivity series on page 45 and name two metals that are more reactive than zinc. Explain your answer.

4 Use the reactivity series on page 45 to decide whether each metal would displace the other metal from its compound:

  a Copper and magnesium chloride.

  b Copper and lead sulphate.

  c Lead and silver nitrate.

  d Iron and copper chloride.

5 Would you expect to find magnesium in the ground as a pure metal element or in an ore? Explain your answer.

6 You have been asked to write a reactivity series for three new metals. Look at the descriptions of the metals below and decide which metal is the most reactive. Which one is the least reactive? Write them in order of their reactivity, with the most reactive metal at the top.

  o Metal A reacts slowly with water, and bubbles steadily when it reacts with an acid.

  o Metal B does not react with oxygen, water or acids.

  o Metal C burns easily in air, and bursts into flames when it reacts with water. It also reacts violently with acids.

Environmental chemistry

What I should already know

- About acids, alkalis, neutralisation reactions and the pH scale.
- Signs that can tell us that a chemical reaction is taking place.
- Examples of chemical reactions in which gases are produced.
- Rocks are broken down into smaller pieces by weathering.
- How plants affect the environment.

What I am going to meet in this unit

- What soil contains.
- Rocks can be weathered by physical, chemical and biological processes.
- How acid rain forms, and its effects.
- Word equations that describe how acid rain damages buildings.
- Monitoring and controlling pollution, and global warming.

What are you doing?

I'm getting ready for global warming!

The **chemical weathering** of rocks is caused by rainwater. **Carbon dioxide** in the air dissolves in rainwater, making it slightly acidic. Sometimes rain can become even more acidic than normal. This **acid rain** weathers rocks and buildings more quickly than normal rain, and can make rivers and lakes so acidic that the plants and animals that live in them die.

As well as carbon dioxide, gases called **sulphur dioxide** and **nitrogen oxides** are produced when we burn fossil fuels. These gases can dissolve in rainwater like carbon dioxide, but when they do they form stronger acids. These acids can be carried for miles before they fall as acid rain.

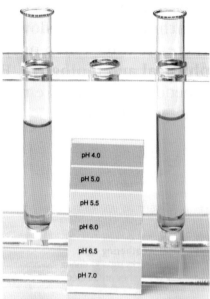

Rainwater is naturally slightly acidic (around pH 6). Acid rain is even more acidic, with a pH of about 4.

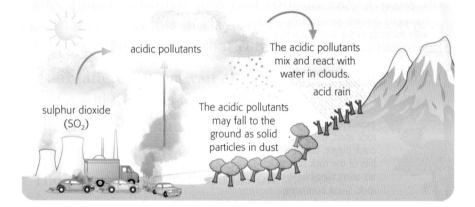

acidic pollutants

The acidic pollutants mix and react with water in clouds.

acid rain

sulphur dioxide (SO₂)

The acidic pollutants may fall to the ground as solid particles in dust

1 What kind of weathering is caused by acid rain?

2 These words have had their vowels removed. What should they say?

   rnwtr   wthrng   fssl fls

3 Acid rain weathers rocks and buildings more quickly than normal rainwater. Explain how acid rain forms. Remember to include the names of the gases that form acid rain.

Copy and complete using these words:

**rainwater   carbon dioxide
chemical   nitrogen oxides**

When we burn fossil fuels, gases called _____ _____, sulphur dioxide and _____ _____ are produced. These can dissolve in _____ to form acid rain, which speeds up the _____ weathering of rocks and buildings.

The effects of acid rain

Acid rain weathers chalk and limestone rocks very quickly. Both of these rocks contain a lot of **calcium carbonate**, which is a metal carbonate. When metal carbonates react with acids a salt, carbon dioxide and water are produced. For example:

calcium carbonate + nitric acid → calcium nitrate + carbon dioxide + water

metal carbonate + acid → salt + carbon dioxide + water

The acids in acid rain also react with **metals** that are used in buildings and sculptures, producing a salt and hydrogen. For example:

iron + sulphuric acid → iron sulphate and hydrogen

metal + acid → salt + hydrogen

This limestone statue is around 500 years old. It has been weathered by acid rain.

As well as affecting buildings and statues, acid rain also affects living things. Acid rain can make lakes and rivers so acidic that no plants or animals can live there. To prevent acid rain from forming, we need to reduce our emissions of sulphur and nitrogen oxides. We can do this by burning less fossil fuels, and using cars that have catalytic converters.

Burning fossil fuels in power stations produces sulphur dioxide.

1 Calcium carbonate is a metal carbonate. Write a word equation to describe what happens when a metal carbonate reacts with an acid.

2 Unscramble these words:

stelam   limcuac bacoanert

3 Describe the effects of acid rain. What can we do to prevent acid rain from forming?

Copy and complete using these words:

living things   metals   acidic calcium carbonate

Acid rain reacts with _____ in buildings and statues, and _____ _____ in rocks. Acid rain also affects _____ _____, by making rivers and lakes too _____ for them to live there.

Global warming

The gases in the Earth's **atmosphere** keep the Earth warm, by acting like the glass in a greenhouse. When sunlight reaches the Earth's surface, infrared radiation is given out. Some of this goes into space, but some is reflected back towards the Earth by **greenhouse gases**. Without this **greenhouse effect**, Earth would be too cold for us to live on.

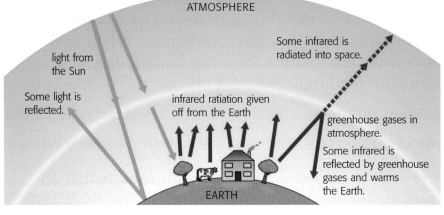

More and more greenhouse gases are being released into the atmosphere. Many people think that this is causing **global warming** (making the Earth warmer). To investigate this theory, scientists collect evidence by measuring changes in water levels, rainfall, plant life and temperature. They can then present their data to the Government, and ask them to make new laws to stop the problem from getting worse.

Greenhouse gas	Produced by
carbon dioxide	burning fossil fuels, respiration and volcanoes
methane	rotting plant and animal material
CFCs	used in fridges and aerosols
nitrogen oxides	traffic pollution

1 What is the greenhouse effect?

2 These words have had their vowels removed. What should they say?

tmsphr   grnhs gss

3 Carbon dioxide is a greenhouse gas. Explain how planting more trees can help to reduce the levels of carbon dioxide in the atmosphere.

Copy and complete using these words:

gases   global warming
temperature   greenhouse effect

The _____ _____ helps to keep the Earth warm. Many people think that our release of greenhouse _____ is causing _____ _____. Scientists investigate this theory by measuring changes in _____ and water levels.

What have I learnt?

1 Some gardeners put compost on their soil. Compost is dead and decayed plant and animal material. Why is this good for the soil?

2 Describe the three different ways by which rocks can be weathered.

3 Name three gases that can dissolve in rainwater to form acid rain.

4 Match up the beginnings and endings below to make complete sentences.

Beginnings

Rainwater is slightly acidic and

Acid rain weathers chalk and limestone quickly

When acid rain reacts with calcium carbonate

When acid rain reacts with some metals

Endings

because they contain lots of calcium carbonate.

a salt and hydrogen are produced.

chemically weathers rocks and buildings.

a salt, carbon dioxide and water are produced.

5 Describe how the greenhouse effect keeps the Earth warm. How are greenhouse gases causing global warming?

6 Write a script for a news report about global warming. Answer these questions in your news report:

o What is global warming?

o How do people think it is caused?

o Does everyone agree about global warming?

o How are scientists investigating global warming?

Using chemistry

What I should already know

- In a chemical reaction, new materials are formed.
- How to write word equations to describe chemical reactions.
- The reactivity series tells us how reactive metals are.
- What displacement reactions are.
- When things get hotter or colder, energy is transferred.

What I am going to meet in this unit

- Burning is a chemical reaction that produces heat energy.
- Some chemical reactions produce heat, light or electrical energy.
- How new materials are designed and produced.
- In a chemical reaction, the total amount of material doesn't change. Mass is conserved.
- We can use the particle model to explain how mass is conserved in a chemical reaction.

Wow. I didn't know that annoying the teacher was an exothermic reaction!

Fuels are substances that release energy when they burn. Burning is also called **combustion**, and happens when a fuel reacts with oxygen. We say that combustion is an **oxidation** reaction.

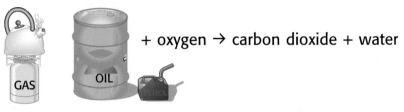

+ oxygen → carbon dioxide + water

hydrocarbons

Many fuels are **hydrocarbons**. Hydrocarbons are compounds that contain carbon and hydrogen only. They produce carbon dioxide and water when they burn completely. This is called **complete combustion**.

For example, natural gas is made of a compound called methane. When it burns:

methane + oxygen → carbon dioxide + water

Sometimes, there is not enough oxygen for a fuel to burn completely. This is called **incomplete combustion**. This wastes the fuel, because not all the energy is released from it. A poisonous gas called carbon monoxide is produced, instead of carbon dioxide.

For lift-off the main engines burn hydrogen. Extra energy comes from burning solid fuel in the white boosters.

1 What is a fuel?

2 What type of reaction is combustion?

3 Propane is a hydrocarbon that is used as a fuel for camping stoves. Write a word equation to describe what happens when propane burns completely.

Copy and complete using these words:
**hydrocarbons   combustion
monoxide   fuels   oxidation   water**

_____ is an _____ reaction that releases energy from _____. Many fuels are _____. They produce carbon dioxide and _____ when they burn completely. Incomplete combustion wastes energy and produces carbon _____.

Heat and light energy are released during the combustion of fuels. This is what makes fuels useful. Reactions that give out heat energy are called **exothermic reactions**.

In a **displacement reaction**, a more reactive metal displaces a less reactive metal from its compound. This releases heat energy.

Displacement reactions can also be used to give out electrical energy. We can make a **voltaic cell** by putting strips of different metals into a salt solution. The further apart the two metals are in the reactivity series, the higher the voltage they produce.

In this skier's hand warmer, two substances react together in an exothermic reaction. The heat that is given out warms her hands.

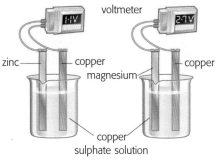

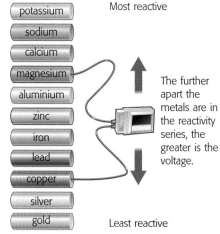

voltmeter

zinc — copper
magnesium — copper

copper sulphate solution

potassium — Most reactive
sodium
calcium
magnesium
aluminium
zinc
iron
lead
copper
silver
gold — Least reactive

The further apart the metals are in the reactivity series, the greater is the voltage.

Magnesium is further away from copper than zinc in the reactivity series, so produces a higher voltage.

1 What do we call a reaction that gives out heat energy?

2 What happens in a displacement reaction?

3 Look at the reactivity series in the picture. Which two metals would produce the highest voltage in a voltaic cell? Explain your answer.

Copy and complete using these words:

**displacement   higher   combustion
further   exothermic   electrical**

_____ reactions like _____ give out heat energy. _____ reactions can be used to make voltaic cells because they give out _____ energy. The _____ apart two metals are in the reactivity series, the _____ the voltage they will produce.

Synthetic materials

Everything around us is made from **materials**: solids, liquids and gases. Some of these are **natural** materials. They are made by reactions in living organisms. Other materials like cosmetics and plastics are made by people in factories. These are called **synthetic** materials. New synthetic materials are made in three stages:

First, people do **research** by asking questions: What properties will the material need? Will people want to buy it? The material is designed and a sample is made.

At the **development** stage, the material is tested to make sure that it has the right properties, and that it is safe. The manufacturer also tries to find a way to make the material as cheaply as possible.

Finally, at the **production** stage, the material is made in a factory. It is then advertised and sold to customers.

1 What do we call materials that are made by people in factories?

2 Unscramble these words:

tentshicy   rulanta   tiermala

3 Draw a flow chart to describe the three stages of making a new material.

Copy and complete using these words:

research   natural   synthetic
production   materials

Everything around us is made of _____. Some of them are _____, but others are _____ and are made by people in factories. The three stages in making a new material are _____, development and _____.

In a chemical reaction, the particles in the reactants rearrange to form the products. No particles are added or lost, so the total **mass** of the products is the same as the total mass of the reactants. We say that mass is **conserved**. This is the law of conservation of mass.

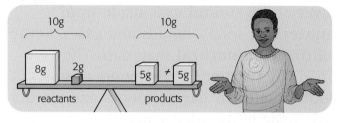

If we have 10 g of reactants, we get 10 g of products because mass is conserved.

For example, if silver nitrate and sodium chloride solutions are mixed, they react to form sodium nitrate and a precipitate of sodium chloride:

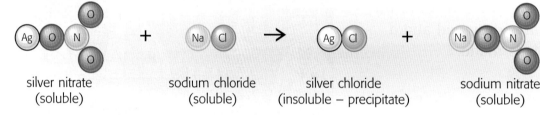

silver nitrate
(soluble)

sodium chloride
(soluble)

silver chloride
(insoluble – precipitate)

sodium nitrate
(soluble)

The number of particles that make up the products is the same as the number of particles that made up the reactants. No particles have been added or lost. This means that the total mass of the products is the same as the total mass of the reactants.

1 What do we mean when we say that mass is conserved in chemical reactions?

2 These words have had their vowels removed, what should they say?

cnsrvd   prtcls   mss

3 60 g of hydrochloric acid reacts with 30 g of calcium carbonate. What are the products of the reaction? What will the total mass of the products be?

Copy and complete using these words:

**mass   products   conserved
total   reactants   particles**

Mass is _____ in chemical reactions. The particles in the _____ rearrange to form the _____. No _____ are added or lost, so the total _____ of the products is the same as the _____ mass of the reactants.

Oxygen and magnesium

When a metal burns in air, a **metal oxide** is produced. For example, magnesium burns in air to produce magnesium oxide:

magnesium + oxygen → magnesium oxide

The magnesium oxide has a greater mass than the magnesium, because the magnesium has joined with oxygen from the air. We can work out how much oxygen reacts with the magnesium because mass is **conserved**:

Weigh a crucible with its lid. Add a piece of magnesium ribbon and weigh again.

Heat crucible with Bunsen burner. Lift lid from time to time to let air in.

Allow apparutus to cool. Reweigh crucible, lid and contents.

In this experiment, 4.8 g of magnesium produced 8.0 g of magnesium oxide. If we take the mass of the magnesium from the mass of the magnesium oxide, we get the mass of the oxygen that was used in the reaction – it was 3.2 g.

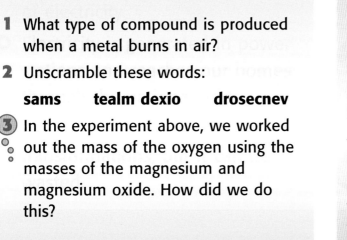

1 What type of compound is produced when a metal burns in air?

2 Unscramble these words:

sams   tealm dexio   drosecnev

3 In the experiment above, we worked out the mass of the oxygen using the masses of the magnesium and magnesium oxide. How did we do this?

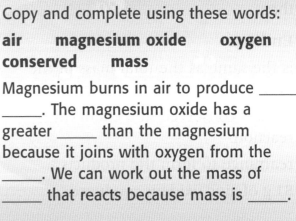

Copy and complete using these words:

**air   magnesium oxide   oxygen
conserved   mass**

Magnesium burns in air to produce _____ _____. The magnesium oxide has a greater _____ than the magnesium because it joins with oxygen from the _____. We can work out the mass of _____ that reacts because mass is _____.

Using energy

Energy is needed to make things happen. There are many different forms of energy. Some of them can be stored. For example, potential (elastic) energy is stored in something that is stretched or twisted. Chemical energy is stored in batteries and food. Kinetic energy is an example of an energy that cannot be stored.

Energy is needed to do work. When work is done, energy is **transferred** (moved from one place to another) and **transformed** (changed from one form to another). For example, when we switch on a lamp, energy is transferred from the wire to the bulb and then to the air around it. The electrical energy is transformed into heat and light energy in the bulb.

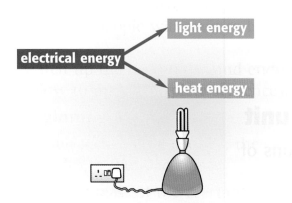

Form of energy	Example
thermal (heat)	
light	
sound	
kinetic (movement)	
potential (gravitational)	
potential (elastic)	
electrical	
chemical	

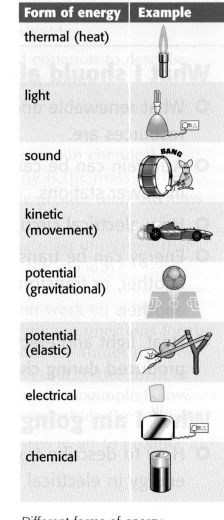

Different forms of energy.

1 What is energy needed for?

2 Name two forms of energy that can be stored, and one form of energy that can't.

3 What is the energy transfer that takes place when we turn on a hairdryer? What are the energy transformations that take place?

Copy and complete using these words:
**transferred   kinetic   forms
energy   chemical   transformed**

_____ is needed to make things happen. When work is done, energy is _____ and _____. There are many different _____ of energy. Some forms, like _____ energy, can be stored but _____ energy can't.

In an electrical circuit, the stored chemical energy in the battery is **transformed** into electrical energy. This is carried around the circuit by **charges**. The electrical energy can then be transformed into other forms of energy if the charges go through an **electrical component** like a bulb.

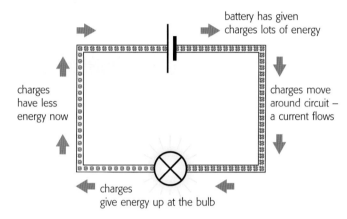

battery has given charges lots of energy

charges have less energy now

charges move around circuit – a current flows

charges give energy up at the bulb

The charges give up their energy as they move round the circuit, but they are not used up.

Current is a measure of the number of charges that pass a certain point in a circuit per second. We measure it in **amperes** (amps or A) using an **ammeter**.

Voltage is a measure of the amount of electrical energy that the charges are carrying around the circuit. We measure it in **volts** (V) using a **voltmeter**.

1 How is electrical energy carried around a circuit?

2 These words have had their vowels removed. What should they say?

crrnt   vltg   chrg

3 Current and voltage are not the same. Explain what current and voltage are, and how they differ. How are current and voltage measured?

Copy and complete using these words:

**voltage   amps   volts   charges
current   electrical**

_____ carry _____ energy around a circuit. _____ is the number of charges passing a certain point in the circuit per second. It is measured in _____. _____ is the amount of energy that the charges are carrying. It is measured in _____.

Electrical energy is generated at power stations. It is **transferred** to our homes through a network of wires called the **national grid**. Once it gets there, it can be transformed into useful forms of energy in **appliances** like TVs and kettles.

Electrical appliances have **power ratings** that tell us how quickly they transform electrical energy. The **power** of an appliance tells us how many joules of energy it uses per second, and is measured in **watts** (W).

$$\text{power (W)} = \frac{\text{energy (J)}}{\text{time (s)}}$$

Electricity companies charge for electrical energy in units called **kilowatt hours** (a 1000 W appliance left on for an hour uses 1 kW of electricity). Appliances that transform electrical energy into heat energy, like hairdryers, usually have higher power ratings than appliances that don't, so they are more expensive to use.

Electricity pylons transfer electricity from power stations to our homes.

| 200 W | 700 W | 2 kW (2000 W) |

Different appliances have different power ratings.

1. What does the power rating of an electrical appliance tell us?

2. Unscramble these words:

   epianplac   eprow   oannialt dirg

3. Look at the appliances above. Which of them will cost the most if it is used for an hour? Explain your answer.

Copy and complete using these words:

joules   power   hours   watts   second   kilowatt

The _____ of an electrical appliance is measured in _____ (W). It tells us how many _____ of electrical energy are used per _____. Electricity companies charge for electrical energy in units called _____ _____.

The chemical energy in fossil fuels, or the kinetic energy of wind or water, is transformed into electrical energy in **power stations**. The transformation happens inside a **generator**, in which a large magnet spins between coils of wire. This makes an electric current flow in the coils, generating electricity.

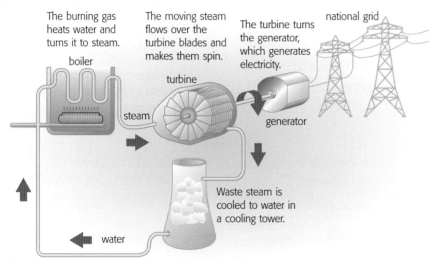

The burning gas heats water and turns it to steam.

The moving steam flows over the turbine blades and makes them spin.

The turbine turns the generator, which generates electricity.

national grid

boiler

turbine

steam

generator

Waste steam is cooled to water in a cooling tower.

water

In a fossil fuel power station, the fuel is burnt to make steam. The steam drives a turbine (a big set of blades) which turns the generator.

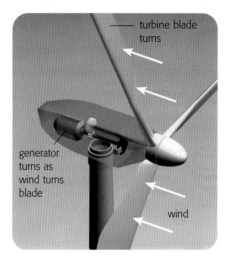

turbine blade turns

generator turns as wind turns blade

wind

This wind turbine is driven by the wind. It turns, then the generator turns and generates electricity.

Burning fossil fuels to generate electricity damages the environment. Greenhouse gases are produced that cause **global warming**. Other gases produced can cause **acid rain**. Renewable energy resources like wind power are better for the environment, and will not run out like fossil fuels.

1 What do generators consist of?

2 These words have had their vowels removed. What should they say?

  glbl wrmng   gnrtr   pwr sttn

3 Give two advantages of using wind power to generate electricity, instead of burning coal.

Copy and complete using these words:

**magnet   current   fossil
generators   fuels   wire**

Electrical energy is generated in power stations in _____. These contain a _____ that spins between coils of _____, generating an electric _____. Burning _____ _____ to generate electricity damages the environment.

When energy is **transformed** (changed from one form to another) there is the same amount of energy at the end as there was at the beginning. This is called the law of conservation of energy. We say that energy is **conserved**.

Although energy is conserved during a transformation, some of it may be transformed into a form that is not useful. We say that this energy is **wasted**. For example, the traditional light bulb in the picture transforms 100 J of electrical energy into 20 J of light energy and 80 J of heat energy. Only the light energy is useful, so the light bulb is only 20% **efficient**.

This hairdryer transforms electrical energy into useful heat energy, and sound energy that is wasted.

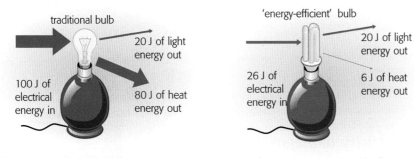

The traditional light bulb is only 20% efficient. The energy-efficient bulb is 77% efficient and wastes a lot less energy.

1 What is the law of conservation of energy?

2 Unscramble these words:

nevoredcs   tificenfe   sedwat

3 A hairdryer transforms 50 J of electrical energy into 30 J of heat energy and 20 J of sound energy. What is the percentage efficiency of the hairdryer?

Copy and complete using these words:

useful   efficient   transformed
conserved   wasted

Energy is _____ when it is _____ from one form into another, but some of it can be transformed into a form of energy that isn't _____. We say that this energy is _____. An appliance that does not waste any energy would be 100% _____.

What have I learnt?

1 Copy and complete these sentences:

a In an electric lamp, electrical energy is transformed into _____ and _____ energy.

b When a drum is hit, _____ (movement) energy is transformed into _____ energy.

c When a vacuum cleaner is switched on, electrical energy is transformed into _____ and _____ energy.

2 Decide whether each statement is true or false. Write them in your book, correcting the ones that are false.

a Electrical energy is measured in watts.

b Power is measured in kilowatt hours.

c Different electrical appliances have different power ratings.

d An electric drill with a power rating of 700 W is more expensive to run than a computer with a 200 W power rating.

3 Sort these sentences into the correct order to explain how electricity is generated in a fossil fuel power station.

- The turbine turns the generator, which generates electricity.
- The heat that is produced turns water into steam.
- The fossil fuel is burnt.
- The electricity goes into a network of wires called the national grid.
- The steam moves over the blades of a turbine and makes them turn.

4 Copy and complete the table using the words below:

Device	Useful energy	Wasted energy
light bulb		
	kinetic	
		sound

heat	heat and sound
hairdryer	light
heat	engine

5 Choose ten important words from this topic and write them in a list. Write a short sentence for each word explaining what it means.

Gravity and space

9J

What I should already know

- Mass is the amount of matter in an object. Weight is a force that pulls objects downwards.

- Why we have day and night, months and seasons.

- The Earth spins on its axis as it orbits the Sun.

- The other planets in the Solar System are different from Earth.

- Natural satellites orbit planets.

What I am going to meet in this unit

- Forces are measured in units called newtons.

- Gravity is a force that pulls together objects that have mass.

- The bigger the mass of two objects and the closer they are, the bigger the force of gravity.

- Different planets have different gravitational field strengths, so your weight would be different on other planets.

- Artificial satellites and their uses.

Gravity pulls things towards the centre of the Earth.

Gravity is a non-contact force. It pulls together objects that have mass. The bigger the mass of the objects, the bigger the **gravitational force** between them. For example, the Earth has a very large mass so objects on its surface or close to it are attracted to it.

Wherever an object is on Earth, it is pulled towards the Earth's centre by gravity.

Wherever you are on Earth, gravity pulls you towards the Earth's centre.

Mass and weight

The **mass** of an object is the amount of matter that it contains, and is measured in **kilograms** (kg). The **weight** of an object is the force that pulls it to the centre of the Earth by gravity, and is measured in **newtons** (N). To work out the weight of an object on Earth you need to multiply the mass by 10. For example, an object that has a mass of 1 kg has a weight of 10 N.

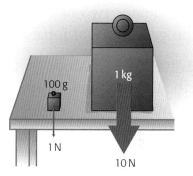

To work out the weight of an object on Earth, multiply the mass (in kg) by 10.

1 What is gravity?

2 Unscramble these words:

   tiwheg   sams   snownet   glakorism

3 A bag of potatoes has a mass of 2 kg. What is its weight on Earth? Is the gravitational force between the bag of potatoes and the Earth greater or less than that between the Earth and a 10 g pencil?

Copy and complete using these words:

kilograms   centre   newtons
mass   gravity   matter

Objects that have _____ are pulled together by a non-contact force called _____. Gravity pulls objects on the surface of the Earth towards its _____. Mass is the amount of _____ that an object contains and is measured in _____. Weight is measured in _____.

The Sun is much larger than the planets in our Solar System. Even though it is a very long way away, the force of gravity is large enough to attract them and keep them in orbit.

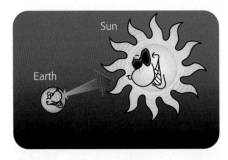

Gravity pulls objects together that have mass. The strength of the **gravitational force** between two objects depends on two things: the **mass** of the objects and the **distance** between them.

o The bigger the mass of two objects, the greater the gravitational force between them.

o The nearer two objects are to each other, the greater the gravitational force between them.

The mass of the Moon is smaller than the mass of the Earth. If you stand on the Moon's surface, the gravitational force between you and the Moon is smaller than that between you and the Earth. This means that if you go to the Moon, you weigh less than you do on Earth, even though your mass hasn't changed.

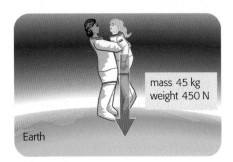

mass 45 kg
weight 450 N

Earth

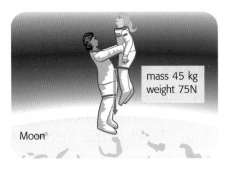

mass 45 kg
weight 75N

Moon

The Moon's mass is smaller than the Earth's, so the force pulling you towards the centre of the Moon is smaller than the force pulling you towards the centre of the Earth.

1 What two things affect the strength of the gravitational force between two objects?

2 These words have had their vowels removed. What should they say?

dstnc   mss   grvttnl frc

3 Would your weight be greater or less than on Earth if you went to the Moon for a holiday? Explain your answer.

Copy and complete using these words:

**nearer   weight   bigger
gravitational   smaller**

The _____ the mass of two objects, and the _____ they are to each other, the greater the _____ force between them. The Moon's mass is _____ than the Earth's. Your _____ would be less on the Moon.

Ideas about the Solar System

We know that the planets in our Solar System orbit the Sun, but people haven't always believed this. People's ideas about the Solar System have developed over time.

In ancient times, people had no scientific evidence to explain events like day and night. They believed that the Earth was flat, and that the planets and stars were gods.

These ideas didn't change until the time of Aristotle (around 350 BC), who suggested that the Sun, planets and the Moon all orbit the Earth. This the **geocentric model** of the Solar System.

Not everyone agreed with the geocentric model. But it was thought to be right until the sixteenth century, when Copernicus suggested that the planets in the Solar System all orbit the Sun. Newton finally proved that this **heliocentric model** was right. He explained how the planets are held in orbit around the Sun by its gravitational force.

1 What were people's ideas about the Solar System in ancient times?

2 Who suggested that the planets in the Solar System orbit the Sun?

3 Draw two diagrams to show the main differences between the geocentric and heliocentric models of the Solar System. Which of the models do we believe today?

Copy and complete using these words:

geocentric   Copernicus   Solar System   Aristotle   heliocentric

People's ideas about the _____ _____ have developed over time. _____ thought that the Sun and planets orbit the Earth. This is the _____ model. Later, _____ suggested that the planets orbit the Sun. This is the _____ model.

We think that the Sun is more than 300 000 times bigger than the Earth. It exerts a huge **gravitational force** on other objects because its mass is so large. Even though it is a very long way away, its gravitational force is large enough to keep the Earth and the other planets in the Solar System in **orbit** around it.

Some of the planets in the Solar System, including the Earth, also have things in orbit around them. Objects that orbit other bodies (planets) are called **satellites**. They are held in orbit by gravity. For example, the Moon is a **natural satellite** that orbits the Earth.

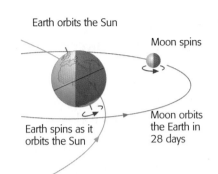

Earth orbits the Sun

Moon spins

Earth spins as it orbits the Sun

Moon orbits the Earth in 28 days

The Earth spins on its axis as it orbits the Sun. The Moon is a natural satellite that orbits the Earth.

Gravity holds planets in orbit around the Sun, like a bucket of water is held in orbit when you swing it round your head on a rope. If the rope breaks, the bucket flies off in a straight line because there is no pulling force to keep the bucket in orbit.

pull of bucket   pull of rope

With no pull from the rope, the bucket flies off in a straight line.

1 Why do Earth and the other planets in the Solar System orbit the Sun?

2 These words have had their vowels removed. What should they say?

   stllt   **rbt**   **ntrl**   **frc**

3 What do we mean when we say that the Moon is a natural satellite of the Earth?

Copy and complete using these words:

natural   mass   orbit   satellites gravitational

The Sun's _____ is very large. The _____ force of the Sun keeps the Earth and the other planets in the Solar System in _____ around it. _____ are objects that orbit bodies like planets. The Moon is a _____ satellite of the Earth.

Artificial satellites are objects that have been put into orbit around the Earth by humans. They are used for different things:

o Communications: sending TV, radio and telephone signals around the world.

o Navigation: The global positioning system (GPS) helps people to work out where they are.

o Observing the Earth and the weather.

o Exploring the Universe.

An artificial satellite in orbit around the Earth.

Satellites can have different types of orbit. **Geostationary** satellites orbit the Earth at the same speed as the Earth's spin, so they stay over the same point. They are used for communications and the global positioning system. **Polar** satellites orbit over the North and South poles.

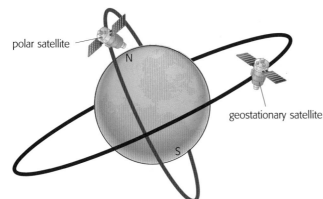

polar satellite

N

S

geostationary satellite

1 What is an artificial satellite? Name two types of artificial satellite orbit.

2 Unscramble these words:

flcairtaii   lrpao   eltisealt

3 What is the global positioning system used for?

Copy and complete using these words:

communications   geostationary
artificial   navigation   polar

_____ satellites are objects that have been put into orbit around the Earth by humans. They are used for _____, _____ and observing the weather. Two types of satellite orbit are _____ and _____.

What have I learnt?

1 Explain the difference between mass and weight. How are mass and weight measured?

2 Jupiter is much bigger than the Earth. Would your weight be bigger or smaller than your weight on Earth if you went to Jupiter for a weekend? Explain your answer.

3 How have people's ideas about the Solar System changed since ancient times? What is the modern-day view of the Solar system?

4 Decide whether each statement is true or false. Write them in your book, correcting the ones that are false.

   a The Sun is three times bigger than the Earth.

   b Gravity holds the planets of the Solar System in orbit around the Sun.

   c Objects which orbit other bodies (planets) are called satellites.

   d The Moon is an artificial satellite of the Earth.

5 Match up the beginnings and endings below to make complete sentences.

Beginnings

Artificial satellites are put into orbit

They can be used for

Two types of satellite orbit are

Geostationary satellites are used for

Endings

polar and geostationary.

communications and navigation.

communications, observing the weather and exploring the universe.

around the Earth by humans.

6 Design a leaflet for next year's Year 9 students about gravity. You should include the answers to these questions in your leaflet:

   o What is gravity?

   o What is the difference between mass and weight?

   o How is gravity affected by mass?

   o How is gravity affected by distance?

   o How would your weight change if you were on a planet that was bigger than the Earth?

   o Would your mass change as well?

   o Why is this?

Speeding up

What I should already know

- Speed tells us how fast something is moving.
- Speed is usually measured in metres per second.
- How to calculate speed.
- Forces affect the speed and direction of objects.
- Friction is a force that stops two surfaces moving over each other.

What I am going to meet in this unit

- How to compare different speeds.
- Acceleration and deceleration.
- If the forces on a stationary object are unbalanced, it will start to move.
- If the forces on a stationary object are balanced, it will not move.
- Streamlining reduces the drag caused by air and water resistance.

Erm... Have you heard about drag?

This is my new design for a sports car. What do you think?

Speed is a measure of how fast something is travelling. When we talk about the speed of something, we usually mean its **average speed**. For example, Jade drives 40 miles to visit her mother. She drives at different speeds during the journey, and sometimes she stops at traffic lights. The whole journey takes her an hour, so her average speed is 40 miles per hour. Her average speed would still be 40 miles per hour if she took two hours to drive 80 miles.

Jade's average speed on the journey to her Mum's is 40 miles per hour.

Different things travel at different speeds. If we know their speeds we can compare them and work out which one is travelling the fastest:

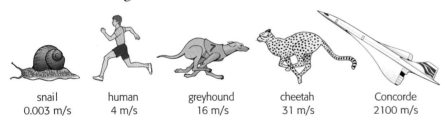

snail	human	greyhound	cheetah	Concorde
0.003 m/s	4 m/s	16 m/s	31 m/s	2100 m/s

1 What is speed?

2 These words have had their vowels removed. What should they say?

vrg   fstst   spd

3 The things in the pictures above are travelling at different speeds. Which of the animals is travelling the fastest? Is this faster or slower than Concorde?

Copy and complete using these words:

**travelling   faster   speed
average   speeds**

_____ is a measure of how fast something is _____. We usually talk about the _____ speed of things. If we know what the _____ of two things are, we can work out which one is travelling _____.

To work out speed of an object, we divide the **distance** that it has travelled by the **time** it took to travel that distance.

For example, a cheetah runs 100 m in 3.22 seconds. What is its average speed?

$$\text{speed} = \frac{\text{distance}}{\text{time}}$$

$$= \frac{100 \text{ m}}{3.22 \text{ s}} = 31.25 \text{ m/s}$$

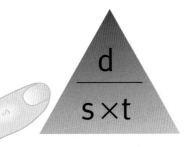

We can use a formula triangle to help us with speed calculations. If you put your finger over the letter that you want to work out (d = distance, t = time and s= speed), it tells you how to do the calculation.

Measuring speed

To calculate speed accurately we need to measure the time carefully. Using a stopwatch is not very accurate because the speed of your reactions can affect the result. In athletics races, photographs are taken of the finishing line to help decide who has won.

This photo shows us who has won the race.

1 What is the formula for calculating speed?

2 If you know the speed of an object and the distance that it has travelled, how could you work out how long it has taken to travel that distance?

3 A car travels 250 miles in 3 hours. What is the average speed of the car? Remember to include the units.

Copy and complete using these words:

**time   distance   carefully
accurately   speed   triangle**

To work out the _____ of the object we divide the _____ that it has travelled by the _____ that it took. A formula _____ can help us with speed calculations. To calculate speed _____ we need to measure time _____.

A **force** is a push or a pull. Forces can alter an object's shape, speed or direction.

If the forces that are acting on an object are **unbalanced**, its speed will change. The object will either **accelerate** (get faster) or **decelerate** (get slower), depending on which force is the biggest.

The forces acting on the skittle are balanced so it does not move.

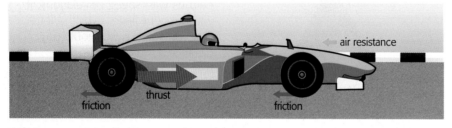

This car is accelerating because the forces acting on it are unbalanced. The forward thrust of the engine is greater than the friction and air resistance.

If an object is stationary and the forces that are acting on it are **balanced**, it will not move. If an object is already moving and the forces that are acting on it are balanced, something different happens. The object does not accelerate or decelerate. It moves at a **constant speed**.

The skater pushes off but then travels at a constant speed. The forces acting on her are balanced.

1 **What are forces? How can they affect objects?**

2 **What do we mean when we say that an object is accelerating?**

3 **You get into a stationary car. Are the forces that act on the car balanced or unbalanced? The car begins to move forwards. Are the forces balanced or unbalanced?**

Copy and complete using these words:

**accelerating   force   speed
decelerating   unbalanced   shape**

A _____ is a push or a pull that can alter an object's _____, _____ or direction. If the forces that are acting on an object are _____ its speed will change. If is gets faster it is _____. If it gets slower it is _____.

Streamlining

Friction, air resistance and water resistance are all forces that act against you to try to slow you down. We call them **drag**.

As you move forward you knock into particles in the air or water. They push against you and slow you down. The faster you travel, the greater the number of particles you collide with and the greater the drag.

Imagine that you are walking along with a big piece of card. If you start to run you can feel the drag increase. This is because more air particles are hitting the card and pushing against you.

One way of reducing the drag that acts against an object is to make it **streamlined** or **aerodynamic**. Streamlined objects are dart or wedge-shaped. They have less surface area for air or water particles to push against when they move, so they can move at greater speeds.

This racing car is streamlined to reduce the effects of drag.

1 What is drag?

2 Unscramble these words:

lastdimener   grad   ria tearnissec

3 The faster you travel, the greater the drag. Explain why this happens. How can we reduce the effects of drag?

Copy and complete using these words:

water resistance   particles
friction   streamlined   drag

_____, air resistance and _____ _____ are forces that cause _____. The faster you travel, the greater the drag because you collide with more air or water _____. We can reduce drag by making objects _____ or aerodynamic.

When a skydiver jumps out of a plane, two forces act on him: **weight** and **air resistance**.

The amount of air resistance depends on his shape and speed:

air resistance

weight

When he jumps from the plane he is not moving very fast. His weight is greater than the air resistance. He accelerates towards the Earth because the two forces are unbalanced.

As he gets faster the air resistance increases. Eventually the two forces are balanced and he moves at a constant speed.

When he opens his parachute the air resistance increases. The forces are now unbalanced and he begins to slow down. This is called deceleration.

He continues to slow down until the air resistance is balanced with his weight. He falls at a constant speed, but more slowly than before.

Once he lands the reaction force of the ground balances his weight.

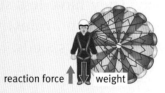

reaction force   weight

1 **What are the forces that act on the skydiver as he falls to Earth?**

2 **These words have had their vowels removed. What should they say?**

   dclrtn   cclrt   wght   blncd

3 Sketch a distance-time graph to show how the skydiver's speed changes during his parachute jump.

Copy and complete using these words:

**unbalanced   shape   accelerates
resistance   speed   weight**

The amount of air resistance that acts on an object is affected by its _____ and _____. Two forces act on a skydiver during a parachute jump: _____ and air _____. When they are _____ the skydiver _____ or decelerates.

What have I learnt?

1 Look at the speeds of the objects below. Which one is travelling the fastest? (Hint: You will need to put the speeds in the same units.)

Object	Speed
A	50 m/s
B	2 km/h
C	20 m/min
D	100 m/s

2 Copy and complete the table below:

Object	Speed	Distance	Time
A		5 km	15 min
B		1.5 miles	30 min
C	1 m/s	2 m	
D	20 m/s		10 s
E		2000 km	2 h

3 Match up the beginnings and endings below to make complete sentences.

Beginnings

Forces can change the

If the forces acting on a stationery object are balanced

If an object's speed increases

If an object's speed decreases

Endings

we say that it is accelerating.

speed, shape or direction of objects.

we say that it is decelerating.

it will not move.

4 Sharks are predators that eat other fish. They have streamlined bodies. How does this help them survive?

5 Write the statements below in the correct order to describe what happens during a parachute jump.

The air resistance increases until it balances with the skydiver's weight. The two forces are balanced.

The skydiver slows down until the forces are balanced. Once he lands, the reaction force of the ground balances his weight.

When the parachute opens the air resistance increases. The forces are not balanced and the skydiver decelerates.

The skydiver's weight is greater than the air resistance. He accelerates towards the Earth.

6 Write an advert for a new sports car. The sports car is very fast because it is streamlined. Draw a picture of the car, and describe how its shape helps it to move quickly. Include these words in your advert:

streamlined   air resistance

drag   air particles   speed

Pressure and moments

9L

What I should already know

- A force is a push or a pull that can change an object's speed, shape or direction.
- Objects accelerate or decelerate when the forces that act on them are unbalanced.
- The role of the skeleton.
- How the particles are arranged in solids, liquids and gases.
- Gas pressure is caused when air particles collide with the sides of a container.

What I am going to meet in this unit

- What pressure is, and how to calculate it.
- Pressure depends on the sizes of the force and the area.
- Hydraulics and pneumatics.
- What levers are and how they work.
- Levers in our bodies.
- Turning effect (moment).

Ever felt under pressure?

Pressure is the effect of a force spread over an area. The amount of pressure depends on the size of the force and the area that it is acting on. The larger the force and the smaller the area it acts on, the greater the pressure.

The snowboard spreads this person's weight so he can glide across the snow.

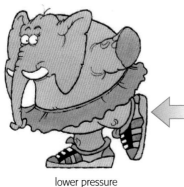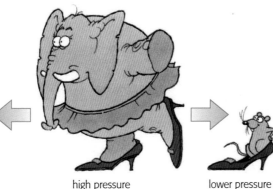

lower pressure   high pressure   lower pressure

We calculate pressure by dividing the **force** (in newtons) by the **area** that it acts on (in square metres).

$$\text{pressure (N/m}^2\text{)} = \frac{\text{force (N)}}{\text{area (m}^2\text{)}}$$

This gives us the pressure in units of **newtons per square metre** (N/m²), which can also be called pascals (Pa).

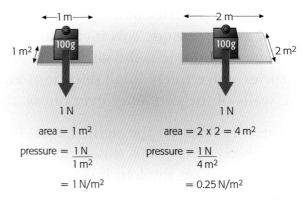

area = 1 m²

pressure = $\frac{1\,N}{1\,m^2}$

= 1 N/m²

area = 2 x 2 = 4 m²

pressure = $\frac{1\,N}{4\,m^2}$

= 0.25 N/m²

These weights have the same mass but they are acting over different areas. The pressure caused by the weight on the right is less because of the bigger area.

1 What is pressure?

2 What is the formula for calculating pressure?

3 A bag weighs 6 N and has an area of 0.5 m². What pressure does the bag exert on the ground? Is this higher or lower than the pressure of a bag that weighs 10 N and has the same area?

Copy and complete using these words:

area   larger   dividing   force
pascals   smaller

Pressure is the effect of a _____ spread over an _____. The _____ the force and the _____ the area, the greater the pressure. We calculate pressure (N/m² or _____) by _____ the force (N) by the area (m²).

Pneumatics

Pneumatics is the study and use of gases under pressure. Gases exert pressure when their particles collide with things, like the walls of a container.

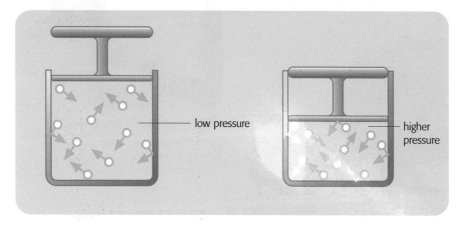

low pressure

higher pressure

Because the particles in a gas are a long way apart, they can be **compressed**. This increases the pressure. In the container on the left, the gas is not compressed. The particles move freely in all directions and collide with the walls of the container every now and then.

In the container on the right the gas has been compressed. It contains the same number of particles as the container on the left, but they are packed closer together. They hit the container sides more often, so the pressure is greater.

1 What is pneumatics?

2 These words have had their vowels removed. What should they say?

cmprssd   prtcl   pnmtcs

3 The containers above contain the same number of gas particles, but the gas pressure is lower on the left than on the right. Explain why.

Copy and complete using these words:

**pressure   pneumatics   particles
compressed   gases**

The study of _____ under pressure is called _____. Gases exert pressure when their _____ collide with things. Because their particles are far apart, they can be _____. This increases the gas _____.

Hydraulics is the study and use of **liquids** to transmit forces by pressure. The particles in a liquid are too close together to be compressed like a gas. Because they can't be compressed, liquids can be used to transmit pressure from one place to another:

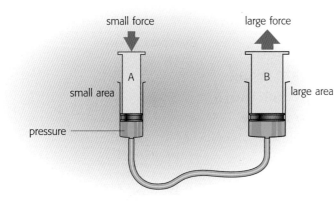

small force

large force

A

B

small area

large area

pressure

The pressure is the same throughout the liquid. Pressing on syringe A produces a larger force at syringe B because it has a larger plunger.

Many hydraulic systems, like car brakes, work like these syringes. The **master piston** acts like the small syringe. Pressure is applied here and transmitted through a liquid to the **slave piston** (acts like the large syringe). This makes the force that acts on the slave piston bigger and the car slows down.

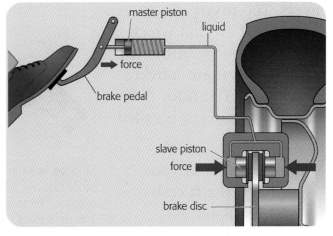

master piston

liquid

force

brake pedal

slave piston

force

brake disc

1 Why can liquids be used to transmit pressure?

2 Unscramble these words:

    luishardyc   quilid   stinop

3 What is the formula for calculating pressure? What happens to the pressure if the area that the force acts on is increased?

Copy and complete using these words:

pressure   hydraulic   particles
liquids   compressed   forces

The use of _____ to transmit _____ by pressure is called hydraulics. Liquids can transmit _____ from one place to another because their _____ are too close together to be _____. Car brakes use a _____ system.

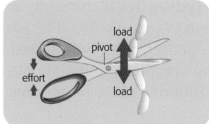

Examples of levers.

Levers make it easier for us to lift things. They do this by **amplifying** (increasing) the force that you are using to move the object. The longer the lever the greater the amplification of the force, and the easier it is to lift the object.

The pictures above show examples of different levers. The **pivot** of a lever is the point around which it moves. The weight of the thing that you are moving is the **load**. The force that you use to move it is called the **effort**.

You have levers in your body that make it easier for you to move.

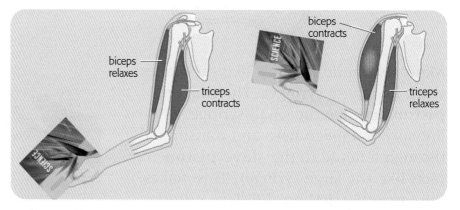

Muscles work in pairs to move a bone at a joint. One contracts whilst the other relaxes. When you lift your arm, your biceps contracts and your triceps relaxes. The small movement of the muscles is amplified by the bone which acts as a lever.

1 What does a lever do?

2 These words have had their vowels removed. What should they say?

mplfng   **ffrt**   **pvt**   **lvr**

3 Crowbars make it easier to remove nails from pieces of wood. Draw and label a diagram to show how a crowbar works as a lever.

Copy and complete using these words:

effort   longer   levers   easier amplifying   pivot

_____ make it easier for us to lift things by _____ the force that you are using. The _____ the lever, the _____ it is to lift the object. The _____ is the point around which the lever moves. The force that you use is called the _____.

The longer the lever, the easier it is to lift something. This is because the longer the lever, the bigger the **turning effect** or **moment**. We can calculate the moment by multiplying the force used (N) by the distance that the effort is from the pivot (m). We measure the moment in units called **newton metres** (N m).

moment = force × distance from pivot

Two loads on a seesaw will only **balance** if the moments are equal on each side. To balance the two moments, we can either change the force or the distance of the load from the pivot.

This seesaw is unbalanced because the moments aren't equal. The force of the elephant is much bigger than that of the sheep.

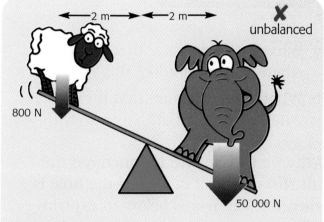

800 N × 2 m = 1600 N m   50 000 N × 2 m = 100 000 N m

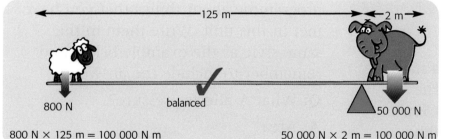

800 N × 125 m = 100 000 N m   50 000 N × 2 m = 100 000 N m

This seesaw is balanced because the moments are equal. The sheep is a long way from the pivot which increases the moment of its force.

1 How do we calculate moment?

2 Unscramble these words:

  alceban   nomtem   nutgirn fectef

3 A caveman uses a lever to lift a large rock. The lever is 2 m long and he applies a force of 100 N. What is the moment? Don't forget to include the units.

Copy and complete using these words:

**bigger   moments   newton
balance   longer   metres**

The _____ the lever, the _____ the moment. Moment is measured in units called _____ _____. Two loads on a seesaw will only _____ if the _____ on both sides are equal.

1 The suitcase below has a weight of 600 N:

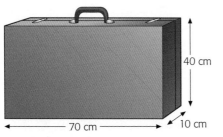

40 cm
70 cm
10 cm

a What is the pressure that it exerts when it is upright like this ?

b What is the pressure that it exerts when it is lying flat?

2 When a bicycle tyre is punctured the air rushes out of it, even if the hole is tiny. Use the words below to explain why:

gas pressure   compressed

particles

3 A force of 500 N is applied to a master piston in a hydraulic system. The master piston has an area of 0.5 m². Calculate the pressure that will be transmitted through the liquid (don't forget to include the units).

4 Your elbow acts like a lever to make it easier for you to move your arm. Draw a diagram of an arm that is lifting something. Label the pivot, the effort and the load.

5 For each of the seesaws below, work out whether or not they would be balanced. State which ends of the unbalanced seesaws would be on the ground.

a Both of the loads have a force of 300 N and are 1 m from the pivot.

b Both of the loads have a force of 600 N. The load on the left is 1 m from the pivot. The load on the right is 2 m away from it.

c The load on the left has a force of 200 N and is 3 m from the pivot. The load on the right has a force of 300 N and is 2 m from the pivot.

6 Imagine that you work in a TV quiz programme. Write ten questions for the programme about things that you have met in this unit. Write them in the same style as the example below and remember to include the answers!

Q: What A affects pressure?

A: Area

Glossary

A

accelerate To increase in *speed* (get faster).

acid A substance with a *pH* lower than 7.

acid rain Rain with a *pH* lower than about 6.

addicted An addicted person keeps wanting to take a *drug*.

addictive An addictive *drug* makes people want to keep taking it.

aerodynamic Another name for *streamlined*.

air resistance A *force* that slows down a falling object.

ammeter An instrument that measures *current*.

ampere (A) Unit of *current*.

amplifying Making bigger. A *lever* amplifies a *force*.

applicance Electrical devices such as kettles and TVs.

area Length × width, usually measured in square metres (m²).

artificial satellite A *satellite* made by people, such as a communications satellite.

artificial selection Another name for *selective breeding*.

asexual reproduction Reproduction with only one parent.

atmosphere The gases that surround the Earth.

average speed The speed over a whole journey. It is total *distance* ÷ total *time*.

B

balanced The same on both sides.

balanced diet A diet with the correct amounts of all the *nutrients* needed for a healthy body.

base A substance that reacts with an *acid* to make a *salt* and *water*.

biological weathering Weathering of rocks caused by plants or animals.

biomass The *mass* of all the *material* in a plant or animal.

C

calcium carbonate A *compound* found in many rocks, which reacts with *acids*.

carbon dioxide A gas found in the air. It is used in *photosynthesis*, and produced in *respiration*.

carnivore An animal that eats only meat.

characteristics Typical features of an organism or object.

charges These carry the *current* around an electrical circuit.

chemical weathering Weathering of rocks caused by chemical reactions, e.g. with *acids*.

chlorides *Salts* which contain chlorine, often formed from hydrochloric acid.

chlorophyll The green chemical in plants needed for *photosynthesis*.

chloroplasts Parts in a plant cell that contain *chlorophyll*. *Photosynthesis* takes place inside the chloroplasts.

circulatory system All the *organs* that work together to transport substances around the body, including the heart and blood vessels.

clone The offspring of *asexual reproduction*, that have the same *genes*.

combustion Another name for burning or *oxidation*.

compete To need the same resources. For example, *weeds* compete with crop plants for light and *water*.

complete combustion Burning in plenty of *oxygen*, so the *fuel* is fully burnt.

compounds Substances made up of more than one *element* combined together.

compressed Squashed together.

conserved Kept the same. *Energy* and mass are both conserved (they are not made or destroyed).

constant speed A *speed* that is steady and does not change.

consumer An animal, that eats plants or other animals.

corrode *Metals* corrode when they are eaten away as they react with *oxygen* and *water*.

crop yield How much of a crop plant is obtained per area of land.

current The flow of electrical *charge* in a circuit.

D

deccelerate To decrease in *speed* (get slower).

deficiency disease A disease caused by not eating enough of a certain *nutrient*.

deforestation Cutting down trees and not replacing them.

desirable characteristics Features that are useful.

development A stage in making a new *material* when the *material* is tested.

diaphragm A sheet of muscle under the lungs which is involved in breathing.

displace To take the place of another *element* in a *compound*.

displacement reaction A reaction in which a more *reactive element* takes the place of a less reactive one in a *compound*.

distance How far it is between one object and another, usually measured in metres.

drag A *force* that slows down a moving object.

drug A substance that affects the way the mind and/or body works.

E

efficient An efficient device *transforms* lots of useful *energy*, and does not waste much energy.

effort The *force* you put in when you use a *lever* or other machine.

egg The female sex cell in animals or plants.

electrical component A device in an electrical circuit, such as a bulb.

element A substance that contains only one type of atom.

energy Energy allows changes to happen.

environmental conditions Things in the surroundings such as light, *water*, and temperature, that affect how well an organism grows.

environmental variation *Variation* from characteristics that are not inherited from parents, but depend on the environmental conditions.

exothermic reaction A reaction which releases *energy*.

F

fertilisation The *gametes* joining together in *sexual reproduction*.

fertilisers Chemicals added to *soil* to provide all the *minerals* for plants to grow well.

fitness How well your body responds to exercise.

food chain A series of organisms linked by food. Each organism is eaten by the next in the chain.

force A push or a pull, measured in *newtons* (*N*).

fossil fuels Coal, oil and natural gas.

fuel A chemical which is burnt to release *energy*.

flower The *organ system* of a plant for *sexual reproduction*.

fruit A part of a plant that contains the *seed*.

G

gametes The cells that join together in *sexual reproduction*.

generator Part of a *power station* where electricity is generated.

genes Genes contain information about the *characteristics* of an *organism*.

geocentric model A model of the Solar System with the Earth at the centre.

geostationary satellite A *satellite* that remains in one position over the Earth's surface.

global warming The idea that the increased *greenhouse effect* is causing the Earth to get warmer.

glucose A sugar which is used in *respiration* and produced in *photosynthesis*.

gravitational force The size of the *force* of *gravity* between two masses.

gravity A *force* that pulls together objects that have mass.

greenhouse A glass building used to grow crop plants.

greenhouse effect *Greenhouse gases* in the *atmosphere* reflect infrared radiation back to Earth, making the Earth warmer.

greenhouse gases Gases in the *atmosphere* that reflect infrared radiation back to Earth.

H

heliocentric model A model of the Solar System with the Sun at the centre.

herbicide A chemical used to kill plants.

herbivore An animal that eats only plants.

humus Dead and decayed plants and animals in *soil*.

hydraulics Using *liquids* to transmit *forces*, using pistons.

hydrocarbons *Compounds* which contain hydrogen and carbon only.

I

illegal drugs Harmful *drugs* taken for recreation, which are against the law.

incomplete combustion Burning in limited *oxygen*, so the fuel is not fully burnt.

inherited variation *Variation* that comes from *characteristics* that are passed from parents to their offspring, in the genes.

K

kilogram (kg) The units of *mass*.

kilowatt hours The units of electrical *energy*.

L

leaf The *organ* of a plant for *photosynthesis*.

lever A simple machine used to lift things more easily.

liquid A state of matter that can flow but cannot be *compressed*.

load The *force* you move when you use a *lever*.

lower epidermis Tissue that covers the underside of a *leaf*.

M

mass The amount of matter that something contains, measured in *kilograms*.

master piston The smaller piston in a *hydraulic* system, that you push.

materials All the substances around us, that we use for different things.

maximum yield The largest amount of crop a farmer can produce from an area of land.

medicinal drugs *Drugs* that are taken to treat illnesses.

metal carbonate A *compound* that may react with an *acid* to produce a *salt, water* and *carbon dioxide*.

metal hydroxide A *compound* that may be formed when a metal reacts with *water*.

metal oxide A *compound* that may be formed when a *metal* reacts with *oxygen*.

metals *Elements* found on the left-hand side of the periodic table. Metals are shiny, hard and feel cold to the touch.

minerals *Compounds* needed for health in animals or plants.

moment The *turning* effect of a *force*, found by *force* × *distance* from *pivot*.

N

national grid Network of wires that carry electricity throughout the country, from *power stations* to homes and factories.

natural *Materials* that occur on Earth, made by living organisms or present in rocks.

natural satellite A body that orbits a planet or star, such as the Moon orbiting Earth.

natural selection Organisms with favourable *characteristics* are well suited to the environment, and they survive and pass them on to the next generation.

neutralisation A reaction between a *base* and an *acid* to form a neutral solution.

newton metres (N m) The units of *turning effect* or *moment*.

newtons (N) The units of *force*.

newtons per square metre (N/m²) The units of *pressure*.

nitrogen A gas in the air. It also occurs in *minerals* that plants need.

nitrogen oxides Gases produced when we burn *fossil fuels*, that cause *acid rain*.

nitrates *Salts* that contain *nitrogen* and *oxygen*, often formed from nitric *acid*.

non-metals *Elements* found on the right-hand side of the periodic table, which do not have the properties of *metals*.

non-selective herbicide A *herbicide* that kills many types of plant.

nutrients Chemicals in foods that we need for a healthy body.

O

obese Very overweight.

omnivore An animal that eats both plants and other animals.

orbit The path of a *satellite* around the object it is travelling around.

organ Part of an organism with a special job, made up of tissues working together.

organ system Many *organs* working together to do a special job.

organic Organic farmers grow crops without using artificial *herbicides* and *pesticides*.

ore A *compound* containing a *metal*. We have to extract the *metal*.

oxidation A reaction in which something reacts with *oxygen*.

oxygen A gas found in the air. It is produced in *photosynthesis*, and used in *respiration*.

P

palisade mesophyll A tissue in a *leaf* made up tall cells with lots of *chloroplasts*.

pattern Something that repeats. For example, lots of similar reactions make up a pattern.

pest An organism that reduces the *crop yield*.

pesticide A chemical used to control or kill *pests*.

pH A scale used to measure *acid*ity and alkalinity. *Acids* have a pH lower than 7. Alkalis have a pH higher than 7.

phloem Tubes that transport dissolved substances around a plant.

phosphorus A *non-metal element*. It occurs in *minerals* that plants need.

photosynthesis Plants making their own food. They change *carbon dioxide* and *water* into *glucose* and *oxygen*, using sunlight and chlorophyll.

physical weathering Breaking up of rocks caused by heating and cooling, or by ice in cracks.

pivot The point around which a *lever* turns.

pneumatics Using gases under *pressure*.

polar satellite A *satellite* whose *orbit* goes over the Earth's North and South Poles.

population A group of organisms of one *species* living in the same habitat.

potassium A *metal element*. It occurs in *minerals* that plants need.

power How much *energy* is used per second, measured in *watts*.

power rating This tells you how quickly an *appliance transforms* electrical *energy*.

power station A place where electrical *energy* is generated.

pressure Force ÷ area, measured in *newtons* per square *metre* (N/m²).

producer A green plant, which produces (makes) it own food.

production A stage in making a new *material* when the *material* is made in a factory.

properties Features of a *material*, such as being shiny or dense.

R

reactive A reactive substance reacts quickly and violently.

reactivity series A list of *element*s (usually *metal*s) in order, from most *reactive* to least reactive.

recreational drugs *Drugs* people take because they enjoy them, such as caffeine.

research A stage in making a new *material* when scientists ask questions, such as what it will be used for and who will buy it.

respiration Living things releasing *energy* from food, using *oxygen*.

respiratory system The organs of the body which work together to provide *oxygen* for *respiration*.

roots The organs of a plant which hold it in the *soil*, and take in *water* and *minerals*.

root hair cells Cells on the roots of plants, which are adapted to absorb *water* and *minerals*.

S

salt A *compound* formed when an *acid* has its hydrogen replaced by a *metal*.

satellite An object that *orbits* another object.

seed The structure formed by *sexual reproduction* that will grow into a new plant.

selected Chosen.

selective breeding Choosing organisms with *desirable characteristics* and mating them to produce offspring that have these *characteristics*.

selective herbicide A *herbicide* that kills only certain types of plant.

sexual reproduction Producing offspring by the fusing of *gametes*.

side effects Unwanted effects of using a *drug*.

slave piston The larger piston in a *hydraulic* system, that provides a larger *force*.

soil A mixture of rock fragments, *humus*, air and *water* in which plants and some animals live.

species Organisms of the same type that can *reproduce* with each other to produce fertile offspring.

speed How fast something is moving. Speed is *distance* moved ÷ *time* taken, often measured in metres per second (m/s).

sperm The male sex cell in animals.

spongy mesophyll A tissue in a *leaf* made up of cells with air spaces between them.

starch A substance used for storing *glucose* in plants.

stem A plant organ that holds the *leaves* and *flowers* above the *soil*.

stoma (plural stomata) A hole in the underside of a *leaf* that lets gases move in and out.

streamlined A shape which reduces *drag*.

sulphates *Salts* which contain sulphur and *oxygen*, often formed from sulphuric acid.

sulphur dioxide Gas produced when we burn *fossil fuels*, that causes *acid rain*.

synthetic Made by people, not found naturally in the environment.

T

tarnish *Metals* tarnish when they go dull as they react with *oxygen* and *water* in the air.

time How long something takes, often measured in seconds (s).

toxins Chemicals which are poisonous.

transfer To move from one place to another.

transferred *Energy* is transferred when it moves from one place to another.

transformed *Energy* is transformed when it changes from one form to another.

turning effect Another name for *moment*.

U

unreactive An unreactive substance reacts slowly or not at all.

unbalanced Not the same on each side.

upper epidermis Tissue that covers the upper surface of a *leaf*.

V

variation Differences between organisms.

vein A tube-like structure in a plant that contains *xylem* and *phloem* tubes.

volt (V) The units of *voltage*.

voltage The amount of electrical *energy* the *charges* are carrying round a circuit.

voltaic cell Strips of different *metal*s in a *salt* solution, that produce a *voltage*.

voltmeter An instrument that measures *voltage*.

W

wasted Not useful. For example, heat *energy* from a bulb is wasted because we want light from a bulb.

water *Compound* that is a *liquid*, involved in many reactions in living things. Water is vital for life.

watts (W) The units of *power*.

waxy cuticle A layer that covers the *upper epidermis* in a leaf, that stops *water* evaporating.

weathered rocks Rocks that have been broken up into tiny fragments.

weed A plant in the wrong place.

weight A *force*. Weight in *newtons* is *mass* in kg × 10.

withdrawal symptoms Bad effects that people experience if they stop taking an *addictive* drug.

X

xylem Tubes that transport *water* around a plant.

Index

95